Current Controversies in Philosophy of Science

Current Controversies in Philosophy of Science asks twelve philosophers to debate six questions that are driving contemporary work in this area of philosophy. The questions are:

I. Are Boltzmann Brains Bad?
II. Does Mathematical Explanation Require Mathematical Truth?
III. Does Quantum Mechanics Suggest Spacetime is Nonfundamental?
IV. Is Evolution Fundamental When It Comes to Defining Biological Ontology?
V. Is Chance Ontologically Fundamental?
VI. Are Sexes Natural Kinds?

These debates explore the philosophical foundations of particular scientific disciplines, while also examining more general issues in the philosophy of science. The result is a book that's perfect for the advanced philosophy student, building up their knowledge of the foundations of the field and engaging with its cutting-edge questions. Preliminary descriptions of each chapter, annotated lists of further readings for each controversy, and study questions for each chapter help provide clearer and richer snapshots of active controversies for all readers.

Shamik Dasgupta is Associate Professor of Philosophy at the University of California, Berkeley, where he has taught since 2016. He works in metaphysics, philosophy of science, and value theory.

Ravit Dotan is a graduate student at the University of California, Berkeley. She specializes in epistemology, philosophy of science, and philosophy of machine learning.

Brad Weslake is Associate Professor of Philosophy at NYU Shanghai. His central research interest is philosophy of science, especially causation and explanation.

Current Controversies in Philosophy

Series Editor: John Turri, *University of Waterloo*

In venerable Socratic fashion, philosophy proceeds best through reasoned conversation. Current Controversies in Philosophy provides short, accessible volumes that cast a spotlight on ongoing central philosophical conversations. In each book, pairs of experts debate four or five key issues of contemporary concern, setting the stage for students, teachers and researchers to join the discussion. Short chapter descriptions precede each chapter, and an annotated bibliography and suggestions for further reading conclude each controversy. In addition, each volume includes both a general introduction and a supplemental guide to further controversies. Combining timely debates with useful pedagogical aids allows the volumes to serve as clear and detailed snapshots, for all levels of readers, of some the most exciting work happening in philosophy today.

Published Volumes in the Series

Current Controversies in Metaphysics
Edited by Elizabeth Barnes

Current Controversies in Values and Science
Edited by Kevin C. Elliott and Daniel Steel

Current Controversies in Philosophy of Religion
Edited by Paul Draper

Current Controversies in Philosophy of Cognitive Science
Edited by Adam J. Lerner, Simon Cullen, and Sarah-Jane Leslie

Current Controversies in Philosophy of Science
Edited by Shamik Dasgupta, Ravit Dotan, and Brad Weslake

For more information about this series, please visit: www.routledge.com/Current-Controversies-in-Philosophy/book-series/CCIP

Current Controversies in Philosophy of Science

Edited by Shamik Dasgupta,
Ravit Dotan, and Brad Weslake

Routledge
Taylor & Francis Group

NEW YORK AND LONDON

First published 2021
by Routledge
52 Vanderbilt Avenue, New York, NY 10017

and by Routledge
2 Park Square, Milton Park, Abingdon, Oxon OX14 4RN

Routledge is an imprint of the Taylor & Francis Group, an informa business

© 2021 Taylor & Francis

Library of Congress Cataloging-in-Publication Data
Names: Dasgupta, Shamik (Writer on Philosophy), editor. |
Dotan, Ravit, editor. | Weslake, Brad, editor.
Title: Current controversies in philosophy of science /
edited by Shamik Dasgupta, Ravit Dotan, and Brad Weslake.
Description: New York, NY : Routledge, 2021. |
Includes bibliographical references and index.
Identifiers: LCCN 2020017026 (print) | LCCN 2020017027 (ebook) |
ISBN 9781138825772 (hardback) | ISBN 9780367531171 (paperback) |
ISBN 9781315713151 (ebook)
Subjects: LCSH: Science–Philosophy.
Classification: LCC Q175 .C943 2020 (print) |
LCC Q175 (ebook) | DDC 501–dc23
LC record available at https://lccn.loc.gov/2020017026
LC ebook record available at https://lccn.loc.gov/2020017027

ISBN: 978-1-138-82577-2 (hbk)
ISBN: 978-1-315-71315-1 (ebk)

Typeset in Bembo
by Newgen Publishing UK

Contents

Contributors

Sean Carroll is Research Professor of Physics at the California Institute of Technology, USA.

Ellen Clarke is Lecturer in Philosophy at the University of Leeds, UK.

Shamik Dasgupta is Associate Professor of Philosophy at the University of California, Berkeley, USA.

Ravit Dotan is a graduate student in the Philosophy Department at the University of California, Berkeley, USA.

Laura Franklin-Hall is Associate Professor of Philosophy at New York University, USA.

Ned Hall is Norman E. Vuilleumier Professor of Philosophy at Harvard University, USA.

J. T. Ismael is Professor of Philosophy at Columbia University, USA.

Muhammad Ali Khalidi is Presidential Professor of Philosophy at City University of New York Graduate Center, USA.

Matthew Kotzen is Associate Professor of Philosophy and Bowman and Gordon Gray Distinguished Term Associate Professor at the University of North Carolina at Chapel Hill, USA.

Mary Leng is Professor of Philosophy at the University of York, UK.

Alyssa Ney is Professor of Philosophy at University of California, Davis, USA.

Maureen A. O'Malley is based in the School of History and Philosophy of Science at the University of Sydney, Australia

Christopher Pincock is Professor of Philosophy at Ohio State University, USA.

David Wallace is Andrew W. Mellon Professor of Philosophy of Science at the University of Pittsburgh, USA.

Brad Weslake is Associate Professor of Philosophy at NYU Shanghai, China.

Introduction

Ravit Dotan and Shamik Dasgupta

What is the philosophy of science? At times, it discusses general questions about the scientific enterprise writ large. For example, science involves *explaining* phenomena, but what is explanation and how does it differ from description and prediction? Observations *confirm* scientific theories, but how can we know which theories are confirmed and to what extent? It is sometimes said that science discovers *laws of nature*, but what are laws and how do they differ from other truths? Science often involves sorting things into *categories*, but does nature itself favor one set of categories or do we impose our own categories on nature? Sometimes scientists conclude that the world is very different from how it appears to us, but what is the relation between the "manifest image" – the world as it appears – and the "scientific image"?

In addition to these general questions, the philosophy of science also discusses more specialized questions with regards to particular scientific disciplines. For example, philosophers of physics ask what we can learn from physics about the structure of space and time, and which interpretation of quantum mechanics is best. Philosophers of biology ask what a species is, and whether there is biological evidence for intelligent design.

On first glance, these two strands to the philosophy of science can appear very different. But on a closer look, the more general questions are often playing out in discussions of the more specialized questions. This book presents six current controversies about more specialized questions that illustrate this. We will describe each controversy in turn.

Are Boltzmann Brains Bad?

As we said, one general issue in the philosophy of science is how evidence confirms scientific theories. The curious case of Boltzmann Brains brings out an interesting puzzle about this. Boltzmann Brains are things that resemble ordinary human brains in all intrinsic respects but are disembodied and came into existence by way of a low-probability thermodynamic fluctuation. Boltzmann Brains have conscious experiences much like ours but are *massively* mistaken about their situation in the world: they think they have bodies and memories of growing up as a child, when in fact they are just an isolated brain that came into existence moments ago. Some cosmological theories imply that the number of Boltzmann Brains in the universe is *vastly* greater than the number of ordinary human beings like us. Sean Carroll argues that such theories are "cognitively unstable": if we gather evidence confirming them we must then conclude that we are probably Boltzmann Brains, but then it follows that our apparent memories of having gathered the evidence are almost certainly false! Hence, argues Carroll, it is impossible to rationally believe the theory. In response, Matthew Kotzen analyzes the notion of cognitive instability in some detail and argues that it is not enough on its own to rule out a theory.

Does Mathematical Explanation Require Mathematical Truth?

Science explains natural phenomena. For example, when two atoms are attracted to one another, we are told, it's because of interactions between the electrons and protons that compose them. When we accept such explanations, we typically expect them to be true – we would expect electrons and protons to actually exist. Can the same be said for mathematical explanations? For example, according to the honeycomb conjecture, honeycombs are made of hexagons because building them with hexagons requires a minimal amount of max. If we accept this explanation, should we take it to be literally true? That is, should we expect whatever mathematical objects it invokes, e.g. geometrical shapes or numbers, to exist? Christopher Pincock gives an affirmative answer. According to Pincock, if mathematical explanations are to genuinely explain anything, they need to be true. Therefore, accepting the honeycomb conjecture entails accepting that mathematical objects exist. Mary Leng disagrees. She develops a fictionalist view on which mathematical explanations can be successful even if they are literally false.

Does Quantum Mechanics Suggest Spacetime Is Nonfundamental?

Physical space appears to be three dimensional, but quantum mechanics calls this apparent truism into question. Certain quantum phenomena suggest that the world we live in is really 3N dimensional, where N is the number of particles in the universe! If so, the three-dimensional space we experience is a massive illusion – or, at best, a derivative phenomenon that ultimately reduces to patterns in a 3N dimensional world. David Wallace and Alyssa Ney discuss this topic, with Ney arguing that this is indeed the correct lesson of quantum mechanics and Wallace arguing that it is not. This debate exemplifies the general question of the relation between the manifest and the scientific image: to what extent can theoretical physics depart from how the world appears and remain empirically adequate?

Is Evolution Fundamental When It Comes to Defining Biological Ontology?

We think of the biological world as composed of individuals. For example, we typically think of a person, a tree, or a fish, as examples of biological individuals. Families, forests, and shoals of fish, on the other hand, are typically thought to be composed of many biological individuals but not themselves individuals. However, it turns out to be surprisingly difficult to draw this distinction when confronted with the rich complexity of biological life. Sometimes, organisms that might ordinarily be counted as distinct turn out to depend on each other in much the way as do cells in a single organism. Should they then count as parts of the same biological individual after all? The general issue playing out in this debate is what determines the correct categories to use in scientific theorizing. Of course, different theories can categorize things differently, but is one of the systems of categorization more fundamental? Ellen Clarke argues that evolutionary theory is fundamental to defining biological individuals, while Maureen O'Malley argues that we can't understand biological individuals without considering other aspects of biological theory.

Is Chance Ontologically Fundamental?

Scientists sometimes collect data about actual events, as when an ecologist counts the number of Lepidoptera butterflies found within a certain area. Other times they look for robust connections between the events, such as *causal* and *counterfactual* relations, or *laws* and *principles* that describe how the events unfold, or what the *chance* of one event is given another. Philosophers of science have long discussed how the robust connections between events relate to the events themselves. On the so-called "Humean" view, the connections are nothing over and above the events. At rock bottom there are just the events; laws, chances, counterfactuals, and so on just reflect patterns in the events. By contrast, the "anti-Humean" view is that the connections cannot be reduced to patterns in the events. Ned Hall and Jenann Ismael discuss this issue as it arises in the case of scientific chances. Ismael defends a broadly Humean account of chance, while Hall argues that Humeanism has counterintuitive implications.

Are Sexes Natural Kinds?

As we mentioned above, one general issue in the philosophy of science is the extent to which the concepts that scientists use in theorizing track natural categories in the world. In their chapters, Muhammad Ali Khalidi and Laura Franklin-Hall discuss this issue as it arises with respect to sex categories throughout the animal kingdom. Are these categories – female and male – imposed by nature, or are they imposed on nature by humans? For example, can these categories be defined across species in terms of relative differences in gamete size? Khalidi argues for an affirmative answer on which sex categories are natural kinds, at least within the animal kingdom. Franklin-Hall disagrees and argues that whether or not sex categories are natural kinds is uncertain.

This list of current controversies is clearly not exhaustive. There are lively contemporary discussions on the role of values in science, the proper use of statistics in scientific reasoning, the relation between science and religion, philosophical issues relating to climate change, and many other topics. We hope this volume inspires you to read more!

Part I

Are Boltzmann Brains Bad?

1 Why Boltzmann Brains Are Bad

Sean M. Carroll

1.1 Introduction

The Boltzmann Brain (BB) problem is a puzzle facing certain kinds of long-lived universes (Dyson, Kleban, & Susskind, 2002; Albrecht & Sorbo, 2004; Page, 2006; Carroll, 2010). In brief, the BB problem arises if our universe (1.1) lasts forever (or at least an extraordinarily long time, much longer than $10^{10^{66}}$ years), and (1.2) undergoes random fluctuations that could potentially create conscious observers. If the rate of fluctuations times the lifetime of the universe is sufficiently large, we would expect a "typical" observer to be such a fluctuation, rather than one of the ordinary observers (OOs) that arise through traditional thermodynamic evolution in the wake of a low-entropy Big Bang. We humans here on Earth have a strong belief that we are OOs, not BBs, so there is apparently something fishy about a cosmological model that predicts that almost all observers are BBs.

This mildly diverting observation becomes more pressing if we notice that the current best-fit model for cosmology – denoted ΛCDM, where Λ stands for the cosmological constant (vacuum energy) and CDM for "cold dark matter" – is arguably a theory that satisfies both conditions (1.1) and (1.2). (Many eternally inflating cosmologies potentially do so as well.) If the vacuum energy remains constant, the universe asymptotically approaches a de Sitter phase. Such a spacetime resembles, in certain ways, a box of gas in equilibrium, with a de Sitter temperature $T_{dS} = \sqrt{\Lambda / 12\pi^2} \sim 10^{-33}\,\text{eV}$. (I will use units where $\hbar = c = k = 1$ unless explicitly indicated.) Since de Sitter space can in principle last forever, thermal fluctuations will arguably give rise to a large number of BBs. In that case, we need to face the prospect that our leading cosmological model, carefully constructed to fit a multitude of astronomical observations and our current understanding of the laws of physics, actually predicts nonsense.

As we will see, the situation is more subtle than this discussion implies; nevertheless, the BB problem is a real one and can act as a constraint on otherwise tenable cosmological models. The real problem is not "this model predicts more BBs than OOs, but we are not BBs, therefore the model is ruled out." The problem is that BB-dominated models are self-undermining, or cognitively unstable – we cannot simultaneously believe that such a model is correct, and believe that we have good reasons for such a belief.

1.2 The Origin of the Problem

To understand the Boltzmann Brain problem, it is useful to revisit the discussions between Ludwig Boltzmann and his contemporaries in the early days of statistical mechanics near the end of the nineteenth century. Boltzmann and others showed how to recover the major results of thermodynamics by considering gases to be large numbers of randomly moving particles. In particular, the Second Law, which states that the entropy S of a closed system will only ever stay constant or increase over time, was shown to be a matter of overwhelming probability rather than absolute certainty.

One way of understanding this result is to use the definition of entropy that is inscribed on Boltzmann's tombstone:

$$S = k \ logW. \tag{1.1}$$

Here k is Boltzmann's constant, and W is the phase-space volume of a macrostate. That is, we imagine taking phase space (the space of all microstates of the theory, in this case the position and momentum of every particle) and partitioning it into macrostates. One common way of doing this is to define some macroscopically observable features of the system, and declaring a macrostate to be the set of all microstates with the same observable properties. Then W can be thought of as the number of microstates in such a macrostate, and the entropy S is the logarithm of this number. It now makes sense why, starting from some low-entropy initial condition, the Second Law is likely to hold: there are many more high-entropy states than low-entropy ones. There will generally be a probability that the entropy of a system will fluctuate downward, from S_1 to $S_1 < S_2$, but that probability will be exponentially suppressed by $\Delta S = S_1 - S_2$:

$$P(\Delta S) \propto e^{-\Delta S}. \tag{1.2}$$

In realistic systems with many particles, this probability becomes negligibly small, so the Second Law is extremely reliable for all practical purposes. (In very small systems with few moving parts, fluctuations can of course be important (Evans & Searle, 2002; Jarzynski, 2011).)

From the very beginning of this program, objections were raised to its fundamental assumptions. One famous objection was based on the recurrence theorem, put forward by Poincaré in 1890. Starting from any initial state and evolving for a sufficiently long time, any classical Hamiltonian system with a bounded phase space will eventually return arbitrarily closely to the original state (Poincaré, 1890). (The quantum version of the theorem posits finite dimensional Hilbert space rather than bounded phase space (Bocchieri & Loinger, 1957; Schulman, 1978; Percival, 1961; Dyson, Kleban, & Susskind, 2002)). Equilibration is not forever, only for a very long time – recurrence times are typically exponential in the maximum entropy of the system. The recurrence theorem implies that eventually, even a system in its highest-entropy configuration will evolve back to a low-entropy initial state, in seeming violation of the Second Law.

Zermelo turned Poincaré's theorem into a full-blown objection to Boltzmann's understanding of the Second Law (Zermelo, 1896/1966a, 1896/1966b). His argument was simple: the recurrence theorem implies that a graph of entropy vs. time must be a periodic function, while the Second Law states that the entropy must monotonically increase, and both cannot be simultaneously true. Zermelo believed that the Second Law was absolute, not merely statistical, and concluded that the mechanistic underpinning to thermodynamics offered by kinetic theory could not be valid.

Boltzmann was convinced that Zermelo's objection was unfounded, but he suggested multiple ways out of the apparent contradiction, and not all of the strategies were mutually consistent (Boltzmann, 1896/1966a, 1897/1966b). (For modern perspectives on the problem, see (Albert, 2003; Carroll 2010).) According to one of his suggestions, perhaps the universe is actually infinitely old, and on very large scales is in thermal equilibrium. Then rare fluctuations would occasionally drive it away from a state of maximal entropy. Boltzmann's twist was to turn lemons into lemonade: maybe we live in the aftermath of just such a fluctuation. (In a slightly earlier paper, he attributes this idea to "my old assistant, Dr. Schuetz" (Boltzmann, 1895).)

In his discussion, Boltzmann anticipates ideas that are often thought of as contemporary and controversial. He imagines a vast cosmos where local conditions are very different in different regions, what a modern physicist would call "the multiverse." He notes that most of this universe is "in thermal equilibrium as a whole and therefore dead," and hence that it is natural we would find ourselves in a relatively atypical region that had departed from equilibrium. This is arguably the first application of the anthropic principle in modern science.

Interestingly, Boltzmann did not seem to recognize an apparently fatal flaw in his scenario. That was left to Eddington, who in 1931 pointed out that anthropic reasoning only requires the existence of intelligent observers, not an entire galaxy such as the one in which we live (Eddington, 1931/2000). According to (1.2), fluctuations into low-entropy states are rare, and fluctuations into even-lower-entropy states are exponentially more rare than that. Consequently, given any particular anthropic criterion, Boltzmann's fluctuation scenario should predict that we live in a universe that has the highest possible entropy, conditioned on that criterion being met. More recently, Albrecht and Sorbo (Albrecht & Sorbo, 2004; Albrecht, 2004), drawing on a discussion by Barrow and Tipler (1986), took this reasoning to its logical conclusion. If our anthropic criterion is simply "the existence of intelligent observers," Boltzmann's fluctuation scenario predicts that the overwhelming majority of such observers should be the absolutely smallest deviations from thermal equilibrium that is compatible with one's definition of an intelligent observer. Plausibly, such a being would require only those body parts that are absolutely essential to cognition and consciousness – it would resemble a disembodied brain surrounded by thermal equilibrium. These minimal observers have become known as Boltzmann Brains (BBs). Boltzmann's fluctuating universe therefore seems to look very different from the relatively orderly cosmos in which we find ourselves.

1.3 What are Boltzmann Brains?

A Boltzmann Brain is a configuration of matter that, along with its local environment, is as close as possible to thermal equilibrium, while still qualifying as an intelligent observer. One might naturally worry that reasoning quantitatively about such a concept would be nearly hopeless, given the difficulty in specifying what it means to be an "intelligent observer." (For an attempt at a careful discussion of what might constitute a BB, and the rate at which they are produced, see De Simone et al., 2010.) In practice, however, such worries turn out not to be very important, due to the kinds of numbers involved. Given any reasonable physical criteria for what constitutes a conscious brain, the ratio of BBs to ordinary observers (OOs) in a given cosmological model is almost always going to be very close to either 0 or incredibly large.

The rate at which BBs fluctuate into existence will be of order $e^{-\Delta S}$, where ΔS is the difference in entropy between the equilibrium state and the state that is almost equilibrium except for a single brain. If we model a region of the universe as a box of gas with a fixed temperature, a typical rate for producing systems with Avogadro's number of protons and neutrons is very roughly

$$\Gamma_{BB} \sim e^{-\Delta S} \sim e^{-10^{66}}. \tag{1.3}$$

There are therefore only really three cases of interest:

- It is possible that the rate of production of BBs is exactly zero. In that case BBs are not a problem.
- The universe could have a finite effective lifetime, e.g. because of a future phase transition or a Big Crunch. In that case it is very easy for the ratio of OOs to BBs to be enormous, just because BBs are so rare.
- The universe could be infinite in duration, and the rate of BB production could be nonzero. In that case BBs will almost always dominate.

Of course it is not only BBs that will fluctuate into existence. We expect not only Boltzmann Brains, but also Boltzmann People, Boltzmann Solar Systems, Boltzmann Galaxies, or even Boltzmann Universes, to eventually appear (Aguirre, Carroll, & Johnson, 2012). Given any particular criterion that we may look for in a fluctuation (such as "the existence of an intelligent observer," or "the existence of a civilization that lasts for at least a million years"), the probability formula (1.2) implies that the most common such fluctuation will be the one with the highest entropy overall. In ordinary, thermodynamically sensible evolution from low-entropy initial conditions, real physical structures often contain substantial mutual information – there are strong correlations between, for example, the state of a camera that has just taken a picture and the state of whatever it was pointed at when the image was taken. There is no reason to expect such correlations in Boltzmann fluctuations, if they are not specifically sought after as part of the criteria defining the fluctuations we are considering. If a camera with an image in its memory (or a person with a memory in their brain) fluctuates into existence, it is overwhelmingly likely that the rest of the universe will still be in thermal equilibrium.

A crucial issue, which we don't have space to properly address here, is determining which cosmological models actually predict that Boltzmann Brains dominate. Conventional wisdom holds that BBs appear when there is a nonzero temperature that persists as the universe expands, a situation that is expected to hold in the presence of a positive vacuum energy (Gibbons & Hawking, 1977). This includes not only speculative scenarios involving inflationary cosmology and the multiverse, but the current stage of our actual universe as well (Riess et al., 1988; Perlmutter et al., 1999). This raises the worry that the real world should be dominated by BBs, independently of any speculative early-universe scenarios.

This conclusion is, at least plausibly, too quick, as argued by Boddy, Carroll, and Pollack (BCP) (2016) (see also Gott, 2008). The quantum vacuum state in a de Sitter spacetime background is a *stationary* state, one that doesn't evolve with time. BCP therefore argued that there aren't true dynamical processes in the de Sitter quantum vacuum state that would produce BBs, so the problem simply doesn't arise. An alternative perspective would be to argue that something "happens" if it occurs within a history that is part of a decoherent set of histories. As Lloyd has argued (2016), it is straightforward to construct sets of decoherent histories in which all sorts of interesting dynamical processes take place, even if the overall density matrix corresponds to a pure stationary quantum state. Taking this as our criteria for the existence of BBs, it follows that BBs are inevitable in almost any cosmological model. Indeed, BBs would occur even in the Minkowski vacuum, with zero cosmological constant and zero temperature; this possibility has been considered (Page, 2006; Davenport & Olum, 2010; Nomura, 2015), but it seems very different from the standard picture of thermal de Sitter fluctuations.

1.4 Why Are Boltzmann Brains Bad?

The standard (but not quite correct) argument that cosmologies dominated by BBs are unacceptable is fairly straightforward: in such a universe, I would probably be a Boltzmann Brain, and I'm not, therefore that's not the universe in which we live. Many people fail to find the standard argument convincing, on the grounds that the existence of observers elsewhere in the universe and very different from themselves shouldn't call into question the success of a model that seems to correctly account for what they see around them (see e.g. Banks, Fischler, & Paban, 2002; Olum, 2002). I will argue that cosmologies dominated by BBs should be rejected, not because I have empirical evidence that I am not one and I should be, but because such models are cognitively unstable.

1.4.1 The Standard Argument

Let's rehearse the standard argument in somewhat more formal terms. Bayes' Theorem relates $P(T_i \mid D)$, the posterior credences we attach to a set of mutually exclusive and exhaustive propositions ("theories") T_i after gathering some data D, to the prior credences $P(T_i)$ and the likelihoods $P(D \mid T_i)$ that such data would be obtained if each proposition were true:

$$P(T_i \mid D) = \frac{P(D \mid T_i) P(T_i)}{P(D)}, \tag{1.4}$$

where the total probability of obtaining the data is $P(D) = \sum_i P(D \mid T_i) P(T_i)$. We would like to use this expression to judge the relative credence we should assign to different cosmological models, where the "data" is taken to be our experience – that we seem to be ordinary observers rather than BBs, or at least that we seem to be embedded in a certain kind of environment that is highly out-of-equilibrium and nicely consistent with thermodynamically conventional evolution from a low-entropy beginning.

Consider two model universes, A with very few BBs, and B in which BBs far outnumber OOs, so that our two propositions are T_A and T_B (we live in universe A or B, respectively). The standard argument assumes that the likelihoods for our observed data in each theory are roughly proportional to the fraction of observers that are OOs:

$$P(D \mid T_i) \sim \frac{N_{OO}}{N_{OO} + N_{BB}(T_i)} \sim \begin{cases} 1 & A, \\ e^{-10^{122}} & B, \end{cases} \tag{1.5}$$

where we have taken $N_{BB}(T_{BB}) \sim \exp(10^{122})$ (a reasonable figure for the measured cosmological parameters of our universe) and we are assuming comparable numbers of ordinary observers in the two theories. This form for the likelihoods is equivalent to an assumption of "the Copernican principle" or "mediocrity" or "typicality" or "self-sampling" or "indifference" concerning who we are in the universe: given some reference class of intelligent observers, we are equally likely to have been any of them, and should reason accordingly (Gott, 1993; Vilenkin, 1995; Bostrom, 2002; Olum, 2002). The numbers in (1.5) are so wildly disparate that, for a very wide range of possible priors, we can immediately conclude to a high degree of confidence that we live in cosmology A, the short-lived one without BBs. This formalizes the intuition that BBs must be subdominant because we are not ourselves BBs.

It's always possible, of course, to choose priors that can overcome even dramatic differences in likelihoods. It might seem fair to assign priors $P(T_A)$ and $P(T_B)$ that are roughly equal to each other, on the theory that we are comparing two cosmological models of comparable plausibility. But one might alternatively adopt a "self-indication assumption" (SIA), according to which we should reason as if we were chosen randomly from the set of all *possible* (rather than actual) observers (Bostrom, 2002; Olum, 2002). Effectively, SIA can be thought of as boosting the prior for theories with BBs by an enormous amount, simply because there are a lot of observers in them:

$$P(T_i) \propto N_{OO} + N_{BB}(T_i). \tag{1.6}$$

This factor would then cancel the suppression in (1.5), undermining the standard argument against BB-dominated cosmologies.

Bostrom (2002) has given an argument against giving higher priors to theories with more observers in them, which he labels the "Presumptuous Philosopher problem." Consider two competing theories that don't have any BBs at all, but that differ substantially in the number of predicted OOs (and are otherwise indistinguishable). Weighting the priors by the number of observers would allow us to essentially rule out − perhaps to extremely high confidence − the universe with fewer observers, even without taking any data or performing any other traditional scientific investigation. The point is that the seemingly innocent and humble assumption that "we are typical observers" isn't innocent at all − it actually grants us enormous leverage in deciding between competing models, by implying that "typical observers are like us." It seems presumptuous to be able to convincingly rule out simple, plausible cosmological scenarios by such nonempirical arguments, just by thinking rather than by collecting data. Let us therefore proceed under the assumption that that the priors for theories A and B are comparable.

1.4.2 Typicality and Likelihoods

The idea that we should reason as if we are typical observers, and therefore that the relevant likelihoods should obey (1.5), has been criticized by Hartle and Srednicki (HS) (Hartle & Srednicki, 2007; Srednicki & Hartle 2010) (also Azhar & Butterfield, 2016; or for an opposing view Bousso, Freivogel, & Yang, 2008). Their argument is closely related to the Presumptuous Philosopher problem. Imagine we learned enough about biochemistry that we were convinced the probability of life coming into existence in the atmosphere of Jupiter was 0.5, and furthermore that if such life did arise, with probability 1 it would evolve an intelligent species with 10^{12} individuals alive today. There are therefore two scenarios in front of us, with equal priors: one in which humans are the only intelligent observers in the Solar System, the other of which over 99% of the intelligent observers are Jovians. HS point out that adopting the likelihoods (1.5), with the role of OOs and BBs replaced by humans and Jovians, respectively, would lead us to conclude that there was less than a one percent chance that the Jovians existed, even though we haven't actually gone to look for them. That's because, in the scenario where there are any Jovians at all, a typical intelligent observer in the Solar System is a Jovian, so the fact that we are human counts as strong evidence against their existence.

Once again, we seem to have helped ourselves to very strong conclusions about the universe without looking at it.

To protect against such presumptuousness, HS argue that a theory of cosmology should be thought of as coming in two parts: a physical model of the universe T_i, which specifies (among other things) the number of any particular kind of observer, and a *xerographic distribution* $\xi_\alpha^{(i)}$, which for each physical model T_i specifies the probability distribution for us to actually be any of the observers in the model. One example of a xerographic distribution would simply be uniform: among all observers, we have an equal probability to be any one of them,

$$\xi_U = \frac{1}{N_{OO} + N_{BB}}. \tag{1.7}$$

Here we have classified every observer as either an OO – defined as observers arising from thermodynamically sensible evolution from very low–entropy conditions, with the ability to draw more or less accurate conclusions about what those conditions were by observing features of their present environment – or a BB (even if the observer is much more than just a brain). But we can also consider distributions that are concentrated entirely on OOs, even in universes that have huge number of BBs in them:

$$\xi_O = \begin{cases} 0 & \text{BBs,} \\ \dfrac{1}{N_{OO}} & \text{OOs.} \end{cases} \tag{1.8}$$

Therefore, we can imagine a cosmological model such as our universe B (which lasts for a recurrence time), but which comes equipped with either a uniform xerographic distribution ξ_U or one ξ_O that is concentrated on OOs. In that case, the likelihood (1.5) corresponding to (B, ξ_U) would be small as before, but the one for (B, ξ_{OO}) would be of order unity. The data that we do not seem to be BBs would count against B with the uniform distribution, but B with a concentrated distribution would be perfectly acceptable, and we would have no right to reject a long–lived cosmological model with dynamical fluctuations (unless the prior were chosen to be very small).

However, we don't quite have data that we are not BBs, or at least not that we aren't Boltzmann Observers who randomly fluctuated into existence in the conditions we seem to find ourselves in. Given only that we are some kind of observer, and a cosmology dominated by random fluctuations, it would indeed be unlikely that we would seem to find ourselves surrounded by a thermodynamically sensible out–of–equilibrium environment; but given that we are exactly the kind of observers we are, in a randomly fluctuating universe it is still overwhelmingly likely that our recent past isn't nearly as low–entropy as we think it is.

A useful approach to the typicality issue that avoids the Presumptuous Philosopher problem, but doesn't sweep the BB problem under the rug, is the suggestion by Neal that we apply "Full Non-indexical Conditioning" (FNC) to our place in the universe (2006). FNC can be summarized by allowing us to use all of the data we currently possess – our personal memories (which might not be accurate, but we have them), the image we perceive of our immediate environments, etc. – but if there are multiple observers within the resulting class, we should reason as if we are chosen from a uniform distribution within that set. Denoting the data available to some particular observer as D_*, this corresponds to choosing a xerographic distribution of the form:

$$
\xi_{FNC} = \begin{cases} 0 & D \neq D_*, \\ \dfrac{1}{N_{D_*}} & D = D_*. \end{cases} \tag{1.9}
$$

Within this scheme, our existence doesn't imply anything about the possibility of many intelligent observers floating in the atmosphere of Jupiter, since our data tells us we are not Jovians.

While there is a superficial resemblance between the distributions (1.8) and (1.9), however, their implications are very different. In a cosmological model of a long-lived universe with random fluctuations, there is unlikely to be more than one OO who has exactly my data (at least within any given Hubble volume), but vast numbers of observers with that data will come from Boltzmann fluctuations. So while conditioning on all the data I have protects me from drawing presumptuous conclusions about life on other planets, it keeps the standard Boltzmann Brain problem completely intact.

The question, then, comes down completely to what kind of priors we should attach to xerographic distributions such as (1.8) and (1.9). There is a crucial philosophical difference between them: the FNC choice (1.9) restricts to a class of observers on the basis of the data they have, while the distribution concentrated on OOs (1.8) references facts that the observers themselves cannot possibly access in their current situation – namely, whether the superficial evidence for a low-entropy past is accurate, or merely a random fluctuation.

That seems like cheating – biasing a xerographic distribution toward a particular answer (we are OOs), not because we have any evidence for it, but because it's the answer we prefer. After all, in a long-lived, randomly fluctuating universe, there would be plenty of observers who arose as random fluctuations rather than from an ultra-low-entropy Big Bang, but who nevertheless drew more or less accurate conclusions about the laws of physics governing their world (even if their number would be dwarfed by the number of true Boltzmann Brains). This would include the kinds of observers Boltzmann himself had in mind, who lived in the aftermath of a fluctuation sufficiently large to make a galaxy like ours, but not necessarily the trillions of galaxies within our observable universe. There would even be many observers who were temporarily fooled, but then successfully updated their models when new data came in (e.g. when the cosmic microwave background suddenly disappeared from their telescopes). Why should we choose a large prior for a xerographic distribution that constitutes an a priori denial that we could possibly be such observers? It seems like we have once again donned the robes of the Presumptuous Philosopher, albeit with a different choice of fabric.

1.4.3 Cognitive Instability

The standard argument against BB-dominated cosmologies is that, in such a cosmology, I should be a BB, but I'm not, therefore that's not the universe I live in. But the analysis above undermines this reasoning somewhat. The caution against presumptuously philosophizing suggests that we should not think of ourselves as being chosen randomly from the set of all intelligent observers in the universe; this gives us too much leverage over what the universe is like, without us actually looking at it. But it also suggests that it's correct to think of ourselves as being chosen randomly from a very restricted class of observers – perhaps as restricted as "observers with precisely my macroscopic data." This prevents us from drawing unwarranted conclusions about other observers in the cosmos, while not sneaking in any unjustified features of our own place within it.

The xerographic distribution (1.9) we are left with, however, doesn't quite do the job of supporting the standard BB argument. While we know (or at least seem to perceive) that we are not disembodied brains floating in an otherwise empty void, we don't know that we and our local environments haven't fluctuated into existence from a higher-entropy equilibrium state – in a long-lived, randomly fluctuating universe, it's overwhelmingly likely that we have indeed done so. We can't claim to have empirical evidence against this disturbing possibility.

What we can do, however, is recognize that it's no way to go through life. The data that an observer just like us has access to includes not only our physical environment, but all of the (purported) memories and knowledge in our brains. In a randomly fluctuating scenario, there's no reason for this "knowledge" to have any correlation whatsoever with the world outside our immediate sensory reach. In particular, it's overwhelmingly likely that everything we think we know about the laws of physics, and the cosmological model we have constructed that predicts we are likely to be random fluctuations, has randomly fluctuated into our heads. There is certainly no reason to trust that our knowledge is accurate, or that we have correctly deduced the predictions of this cosmological model.

A randomly fluctuating Boltzmann universe puts us, then, in a strange predicament. On the one hand, we use our reasoning skills and knowledge of physics to deduce that in such a cosmos we are probably randomly fluctuated observers, even after conditioning on our local data. On the other hand, we can also deduce that we then have no reason to trust those reasoning skills or that knowledge of physics.

The randomly fluctuating universe scenario is therefore self-undermining, or as Albert has characterized similar situations in statistical mechanics, *cognitively unstable* (Albert, 2003; Carroll, 2010; Myrvold, 2016). If you reason yourself into believing that you live in such a universe, you have to conclude that you have no justification for accepting your own reasoning. You cannot simultaneously conclude that you live in a randomly fluctuating universe and believe that you have good reason for concluding that. In (what most of us believe to be) the real world, in which there really was a low-entropy Big Bang in the relatively recent past, our memories and deductions about the past rely on a low-entropy Past Hypothesis for their justification. In a universe dominated by Boltzmann fluctuations, such a hypothesis is lacking, and we can't trust anything we think we know.

How are we to treat the prospect of a cognitively unstable theory? We can't say for sure that it is not true – it's *possible* that we do live in a fluctuation away from equilibrium, perhaps with completely false memories of the past. We can't even collect evidence that would noticeably decrease our credence in the possibility, as such evidence is also likely to have fluctuated into our heads.

The best we can do is to decline to entertain the possibility that the universe is described by a cognitively unstable theory, by setting our prior for such a possibility to zero (or at least very close to it). That is what priors are all about: setting credences for models on the basis of how simple and reasonable they seem to be before we have collected any relevant data. It seems unreasonable to grant substantial credence to the prospect that we have no right to be granting substantial credence to anything. If we discover that a certain otherwise innocuous cosmological model doesn't allow us to have a reasonable degree of confidence in science and the empirical method, it makes sense to reject that model, if only on pragmatic grounds. This includes theories in which the universe is dominated by Boltzmann Brains and other random fluctuations. It's not that we've gathered evidence against such theories by noticing that we are not BBs; it's that we should discard such theories from consideration even before we've looked.

It's important that we're not merely discarding the possibility that we ourselves are BBs; we are discarding the cosmological models in which they dominate. We might ask whether this kind of reasoning should simply lead us to reject xerographic distributions that allow us to be observers whose memories are not accurate – the universe is full of BBs, but we're not among them. But as discussed above, this represents a criterion based on facts about our location in the universe that we ourselves do not have access to, which raises immediate problems. As a matter of empirical observation, there are many people around us who have wildly inaccurate memories or understandings of the laws of physics; it would seem dubious to exclude them from our xerographic distribution. In a long-lived, randomly fluctuating universe, the dividing line between observers that have been fooled by statistically unlikely data and those than have simply drawn incorrect conclusions from the data they've collected is a hard one to draw. And many observers with data just like ours may have been fooled thus far, but will eventually figure things out when they realize that the microwave background has disappeared. We are on much safer footing when we choose xerographic distributions that discriminate on the basis of the currently observed situations that observers are in, rather than the accuracy of their deductions or the sensibility of their thermodynamic evolutions.

1.5 Conclusions

We therefore conclude that the right strategy is to reject cosmological models that would be dominated by Boltzmann Brains (or at least Boltzmann Observers among those who have data just like ours), not because we have empirical evidence against them, but because they are cognitively unstable and therefore self-undermining and unworthy of serious consideration. If we construct a model such as ΛCDM or a particular instantiation of the inflationary multiverse that seems to lead us into such a situation, our job as cosmologists is to modify it until this problem is solved, or search for a better theory. This is very useful guidance when it comes to the difficult task of building theories that describe the universe as a whole.

Acknowledgements

I have benefited from discussions with a large number of people over the years, including David Albert, Andy Albrecht, Anthony Aguirre, Charlie Bennett, Kim Boddy, Raphael Bousso, Jennifer Chen, Alan Guth, Jim Hartle, Matt Johnson, Andrei Linde, Don Page, Jason Pollack, Jess Riedel, and Mark Srednicki. This research is funded in part by the Walter Burke Institute for Theoretical Physics at Caltech, by DOE grant DE-SC0011632, by the Foundational Questions Institute, and by the Gordon and Betty Moore Foundation through Grant 776 to the Caltech Moore Center for Theoretical Cosmology and Physics.

References

Aguirre, Anthony, Carroll, Sean M., & Johnson, Matthew C. (2012). Out of equilibrium: understanding cosmological evolution to lower-entropy states. *Journal of Cosmology and Astroparticle Physics*, 2, 24. [arXiv:1108.0417 [hep-th]]

Albert, David Z. (2000). *Time and chance*. Cambridge, MA: Harvard University Press.

Albrecht, Andreas. (2004). Cosmic inflation and the arrow of time. In John B. Barrow, Paul W. C. Davies, & Charles L. Harper Jr. (Eds.), *Science and ultimate reality* (pp. 363–401). Cambridge, UK: Cambridge University Press. [arXiv:astro-ph/0210527]

Albrecht, Andreas, & Sorbo, Lorenzo. (2004). Can the universe afford inflation? *Physical Review D*, 70(6). [arXiv:hep-th/0405270]

Azhar, Feraz, & Butterfield, Jeremy. (2016). Scientific realism and primordial cosmology. [ArXiv preprint at arXiv:1606.04071 [physics.hist-ph]]

Banks, Tom, Fischler, Willy, & Paban, Sonia. (2002). Recurrent nightmares?: measurement theory in de Sitter space. *Journal of High Energy Physics*, 12, 062. [arXiv: hep-th/0210160]

Barrow, John D., & Tipler, Frank J. (1986). *The anthropic cosmological principle*. New York, NY: Oxford University Press.

Bocchieri, P., & Loinger, A. (1957). Quantum recurrence theorem. *Physical Review*, 107(2), 337–338.

Boddy, Kimberly K., Carroll, Sean M., & Pollack, Jason. (2017). Why Boltzmann brains don't fluctuate into existence from the De Sitter vacuum. In Khalil Chamcham, Joseph Silk, John D. Barrow, et al. (Eds.), The philosophy of cosmology (pp. 228–240). Cambridge, UK: Cambridge University Press. [arXiv:1505.02780 [hep-th]

Boddy, Kimberly K., Carroll, Sean M., & Pollack, Jason. (2016). De Sitter space without dynamical quantum fluctuations. *Foundations of Physics, 46*(6), 702–735. [arXiv:1405.0298 [hep-th]]

Boltzmann, Ludwig. (1895). On certain questions of the theory of gases. *Nature, 51,* 413–415.

Boltzmann, Ludwig. (1966a). Entgegnung auf die wärmetheoretischen Betrachtungen des Hrn. E. Zermelo. In Stephen G. Brush (Trans.), *Kinetic Theory* (p. 392). Oxford, UK: Oxford University Press. [Originally published in 1896, in *Annalen der Physik, 293*(4), 773–784]

Boltzmann, Ludwig. (1966b). Zu Hrn. Zermelo's Abhandlung "Ueber die mechanische Erklärung irreversibler Vorgänge." In Stephen G. Brush (Trans.), *Kinetic Theory* (p. 412). Oxford, UK: Oxford University Press. [Originally published in 1897, in *Annalen der Physik, 296*(2), 392–398]

Bostrom, Nick. (2002). *Anthropic bias: observation selection effects in science and philosophy.* New York: Routledge.

Bousso, Raphael, Freivogel, Ben, & Yang, I. Sheng. (2008). Boltzmann babies in the proper time measure. *Physical Review D, 77*(10), 103514. [arXiv:0712.3324 [hep-th]]

Carroll, Sean M. (2010). *From eternity to here: the quest for the ultimate theory of time.* New York, NY: Penguin.

Davenport, Matthew, & Olum, Ken D. (2010). Are there Boltzmann brains in the vacuum? [ArXiv preprint at arXiv:1008.0808 [hep-th]]

De Simone, Andrea, Guth, Alan H., Linde, Andrei, et al. (2010). Boltzmann brains and the scale-factor cutoff measure of the multiverse. *Physical Review D, 82,* 063520. [arXiv:0808.3778 [hep-th]]

Dyson, Lisa, Kleban, Matthew, & Susskind, Leonard. (2002). Disturbing implications of a cosmological constant. *Journal of High Energy Physics, 10,* 011. [arXiv:hep-th/0208013]

Eddington, Arthur S. (2000). Driven to admit anti-chance. In D. R. Danielson (Ed.), *The Book of the Cosmos: imagining the Universe from Heraclitus to Hawking* (pp. 401–406). Cambridge, MA: Perseus. [Originally published in 1931, in *Nature, 127,* 3203]

Evans, Denis J., & Searles, Debra J. (2002). The fluctuation theorem. *Advances in Physics, 51*(7), 1529–1585.

Gibbons, Gary W., & Hawking, Stephen W. (1977). Action integrals and partition functions in quantum gravity. *Physical Review D, 15*(10), 2752.

Goldstein, Sheldon, Struyve, Ward, & Tumulka, Roderich. (2015). The Bohmian approach to the problems of cosmological quantum fluctuations. [ArXiv preprint at arXiv:1508.01017 [gr-qc]]

Gott, J. Richard, III. (1993). Implications of the Copernican principle for our future prospects. *Nature, 363,* 315–319.

Gott, J. Richard, III. (2008). Boltzmann brains – I'd rather see than be one. [ArXiv preprint at arXiv:0802.0233 [gr-qc]]

Hartle, James B., & Srednicki, Mark. (2007). Are we typical? *Physical Review D, 75*(12), 123523. [arXiv:0704.2630 [hep-th]]

Jarzynski, Christopher. (2011). Equalities and inequalities: irreversibility and the second law of thermodynamics at the anoscale. *Annual Review of Condensed Matter Physics, 2*(1): 329–351.

Lloyd, Seth. (2016). Decoherent histories approach to the cosmological measure problem. [ArXiv preprint at arXiv:1608.05672 [quant-ph]]

Myrvold, Wayne C. (2016). Probabilities in statistical mechanics. In Alan Hajek & Christopher Hitchcock (Eds.), *The Oxford Handbook of Probability and Philosophy* (pp. 573–600). Oxford: Oxford University Press.

Neal, Radford M. (2006). Puzzles of anthropic reasoning resolved using full non-indexical conditioning. *ArXiv Preprint Math/0608592*. [ArXiv preprint at arXiv:math/0608592 [math.ST]]

Nomura, Yasunori. (2015). A note on Boltzmann Brains. *Physics Letters B, 749*, 514–518. [arXiv:1502.05401 [hep-th]]

Olum, Ken D. (2002). The doomsday argument and the number of possible observers. *The Philosophical Quarterly, 52*(207), 164–184. [arXiv:gr-qc/0009081]

Page, Don N. (2006). The lifetime of the universe. *Journal of the Korean Physical Society, 49*, 711–714. [arXiv: hep-th/0510003]

Percival, Ian C. (1961). Almost periodicity and the quantal H theorem. *Journal of Mathematical Physics, 2*(2), 235–239.

Perlmutter, S., Aldering, G., Goldhaber, G., et al. (1999). Measurements of omega and lambda from 42 high-redshift supernovae. *The Astronomical Journal, 517*(2), 565–586. [arXiv: astro-ph/9812133]

Poincaré, Henri. (1890). Sur le problème des trois corps et les équations de la dynamique. *Acta Mathematica, 13*, 1–270.

Riess, Adam G., Filippenko, Alexei V., Challis, Peter, et al. (1998). Observational evidence from supernovae for an accelerating universe and a cosmological constant. *The Astronomical Journal, 116*, 1009. [arXiv:astro-ph/9805201]

Schulman, Lawrence .. S. (1978). Note on the quantum recurrence theorem. *Physical Review A, 18*(5), 2379–2380.

Srednicki, Mark, & Hartle, James. (2010). Science in a very large universe. *Physical Review D, 81*(12), 123524. [arXiv:0906.0042 [hep-th]]

Vilenkin, Alexander. (1995). Predictions from quantum cosmology. *Physical Review Letters, 74*(6), 846. [arXiv:gr-qc/9406010]

Zermelo, Ernst. (1966a). Über einen Satz der Dynamik und die mechanische Wärmetheorie. In Stephen. G. Brush (Trans.), *Kinetic Theory* (p 382). Oxford, UK: Oxford University Press. [Originally published in 1896, in *Annalen der Physik, 293*(3), 485–494]

Zermelo, Ernst. (1966b). Über mechanische Erklärungen irreversibler Vorgänge. Eine Antwort auf Hrn. Boltzmann's „Entgegnung." In Stephen G. Brush (Trans.), *Kinetic Theory* (p. 403). Oxford, UK: Oxford University Press. [Originally published in 1896, in *Annalen der Physik, 295*(12), 793–801]

2 What Follows from the Possibility of Boltzmann Brains?

Matthew Kotzen

2.1 The Boltzmann Brain Problems

A Boltzmann Brain is a hypothesized observer that comes into existence by way of an extremely low-probability quantum or thermodynamic[1] "fluctuation" and that is capable of conscious experience (including sensory experience and apparent memories) and at least some degree of reflection about itself and its environment. Boltzmann Brains do not have histories that are anything like the ones that we seriously consider as candidates for own history; they did not come into existence on a large, stable planet, and their existence is not the result of any sort of evolutionary process or intelligent design. Rather, they are staggeringly improbable cosmic "accidents" that are (at least typically) massively deluded about their own predicament and history. It is uncontroversial that Boltzmann Brains are both metaphysically and physically possible, and yet that they are staggeringly unlikely to fluctuate into existence at any particular moment.[2] Throughout the following, I will use the term "ordinary observer" to refer to an observer who is not a Boltzmann Brain. We naturally take ourselves to be ordinary observers, and I will not be arguing that we are in any way wrong to do so.

There are several deep and fascinating philosophical and cosmological questions that are raised by the possibility of Boltzmann Brains. Here are just a few of them: Do I have compelling reasons to believe that I am not a Boltzmann Brain? Is it possible, under any circumstances, for me to coherently believe that I am a Boltzmann Brain, or to suspect that I might be? If I am persuaded that most of the observers in the universe are indeed Boltzmann Brains, does that rationally compel me to believe that I am most likely a Boltzmann Brain? If I am persuaded that most of the observers in the universe *who are in my subjective state* are Boltzmann Brains, does that rationally compel me to believe that I am most likely a Boltzmann Brain? If a given cosmology entails that I am most likely a Boltzmann Brain, does this provide a strong reason to reject that cosmology? Does a cosmology according to which the universe has an infinite past or future (or both), or infinite space, entail that I am most likely a Boltzmann Brain — and, if so, should such a cosmology for this reason be rejected? Does a cosmology according to which the entire universe, or the portion of the universe in which we live, is the result of a quantum or thermodynamic fluctuation entail that I am most likely a Boltzmann Brain — and, if so, should such a cosmology for this reason be rejected? Are there available and plausible and evidentially supported cosmologies according to which I am most likely *not* a Boltzmann Brain – and, if so, does this provide an additional reason to accept such cosmologies? Does the possibility of Boltzmann Brains present a genuine philosophical paradox, where several independently compelling hypotheses turn out to be logically inconsistent with each other, or does it simply present problems for particular cosmological assumptions and not others?

I cannot hope to resolve – or even to substantially address – all of these large and difficult questions here. However, in this chapter, I will try to do three things. First, in Section 2.2, I will argue that though David Albert and Sean Carroll's notion of "cognitive instability" is an interesting and important one, considerations of cognitive instability alone are not sufficient to rule out the hypothesis that I am a Boltzmann Brain. In Section 2.3, I will argue against James Hartle and Mark Srednicki's conclusion that we can coherently believe both that most observers in the universe are Boltzmann Brains and yet that we (likely) aren't. And in Section 2.4, I will briefly survey the nature of the problems that Boltzmann Brains pose for several categories of cosmological hypotheses.

2.2 Cognitive Instability

Carroll, building on some suggestions by Albert, has argued that I should reject the hypothesis that I am a Boltzmann Brain because there is an important sense in which belief in such a hypothesis is "cognitively unstable" and hence self-undermining. Carroll writes:

> Is it possible that you and your surrounding environment, including all of your purported knowledge of the past and the outside world, randomly fluctuated into existence out of a chaotic soup of particles? Sure, it's possible. But you should never attach very high credence to the possibility. Such a scenario is *cognitively unstable*, in the words of David Albert. You use your hard-won scientific knowledge to put together a picture of the world, and you realize that in that picture it is overwhelmingly likely that you have just randomly fluctuated into existence. But in that case, your hard-won scientific knowledge just randomly fluctuated into existence as well; you have no reason to actually think that it represents an accurate view of reality. It is impossible for a scenario like this to be true and at the same time for us to have good reasons to believe in it. The best response is to assign it a very low credence and move on with our lives.
>
> (2016, p. 92)

There is an ambiguity here: is Carroll's conclusion merely that we should *never* attach a *very high* credence to the possibility that we are Boltzmann Brains, or is it that we *should* attach a *very low* credence to the possibility? After all, there is significant credal distance between "very high" and "very low." Some version of the former conclusion may be plausible, but I am not persuaded of the latter conclusion. After all, there are plenty of other widely discussed "skeptical" hypotheses that also, arguably, exhibit the phenomenon of cognitive instability, but I do not think that this alone provides adequate reason for us to reject (or to assign very low credence to) those hypotheses.

For instance, Descartes' famous "Dreaming Hypothesis" – the hypothesis that I'm currently experiencing a lifelike dream as opposed to interacting directly with the external world – may exhibit some degree of cognitive instability. If I were to believe that I am currently dreaming on the basis of my available evidence, there is a threat that the significance of that evidence is significantly undermined by the Dreaming Hypothesis itself; if I am dreaming right now, then (at least much of) my available evidence is dreamt evidence, and dreamt evidence (much like randomly fluctuated evidence) also fails to represent an accurate view of reality. Similar considerations apply to more contemporary skeptical hypothesis such as the hypothesis that all of my experiences are being caused by electrical stimulation via a supercomputer such as The Matrix; experiences caused by electrical stimulation via The Matrix similarly fail to represent an accurate view of reality. But surely this observation alone is not a general solution to Cartesian skepticism; even if I cannot reasonably assign very high credence to the Dreaming Hypothesis on the basis of evidence which would be unreliable if the Dreaming Hypothesis is true, it doesn't follow that I should assign a very low credence to the Dreaming Hypothesis. Rather, perhaps I just have no rational basis for determining whether I am dreaming or not, in which case one very natural response seems to be to assign some sort of middling credence – neither very high nor very low – to the Dreaming Hypothesis.[3]

Another example of cognitively unstable hypotheses is a class of "conspiracy theories" according to which a powerful agent or group has endeavored to mislead the public about their activities. If I come to believe that such a powerful agent is manipulating all of my evidence, including my evidence about the evidence-manipulating activities of that very agent, then it is hard to see how I could coherently hold an evidence-based belief in the truth of the conspiracy theory in question. But again, it is not obvious that the best response in each case is to simply assign a very low credence to the conspiracy theory in question; if the theory is independently plausible and does a good job of explaining the available evidence, then it may well deserve to be taken seriously and to be assigned a credence that is higher than "very low."

Though introduced for largely distinct dialectical purposes, Adam Elga's example of hypoxia (a condition in which the body and brain are deprived of adequate oxygen) also raises similar issues:

> Hypoxia impairs reasoning and judgment, which is bad enough. But what makes the condition really insidious is that it can be undetectable. In other words, when hypoxia impairs someone's reasoning, it can do so in such a way that the impaired reasoning seems (to the hypoxic individual) to be perfectly fine. It is a sad truth that airline pilots have crashed planes by happily and confidently making horrible judgment calls as a result of suffering from hypoxia.
>
> (Elga, 2008, p. 3)

Is it possible for the hypothesis that I'm currently hypoxic to be true and at the same time for me to have good reasons to believe it? The answer to this question may well depend on some of the details of the effects of hypoxia. Does hypoxia cause people to reason poorly in general, including when they are performing basic deductive and inductive inferences, or are the effects restricted to "higher-level" inferential skills such as those deployed in complicated mathematical calculations, strategic reasoning, and spatial reasoning? If hypoxia's effects are limited to higher-level forms of reasoning, then a hypoxic person might well double-check her mathematical calculations with a calculator, or measure the oxygen level in her own blood, and thereby come to the reasonable conclusion that she is hypoxic. But if hypoxia impacts basic reasoning, then it seems that almost *any* conclusion that a hypoxic person draws about her hypoxia – including the conclusion that she is hypoxic – is bound to be an unreasonable one. The conclusion in question might well seem like a reasonable one to the hypoxic person given her evidence, of course, but that's just a symptom of the hypoxia. If this is how hypoxia works, then it is hard to see how it could be possible for the hypothesis that a particular person is hypoxic to be true and at the same time for her to have a justified belief that it is true. However, the lesson here cannot be that the appropriate response is for pilots to simply disregard or discount the hypothesis that they are hypoxic. The "sad truth" that Elga refers to is that pilots suffering from hypoxia often assign *far too low* a credence to the hypothesis that they are hypoxic, and – notwithstanding the cognitive instability of the hypothesis that an individual is hypoxic – we certainly wouldn't want to instruct new pilots to adopt a policy of assigning a very low credence to the hypothesis that they are hypoxic and (hope that they) move on with their lives.

Some philosophers might try to distinguish examples like hypoxia from the case of Boltzmann Brains by arguing that whereas hypoxia makes it impossible for me to *take appropriate account of* the evidence I have that I'm hypoxic, I still *have* the evidence that I am hypoxic (say, in the form of my poor decision-making), right there in front of me; by contrast, since (as Carroll argues) fluctuated evidence doesn't actually represent an accurate view of reality, it is impossible for me to even *have* good reasons to believe that I am Boltzmann Brain.[4] In the philosophical jargon, another way to put this point might be that the truth of the hypoxia hypothesis merely makes it impossible for me to have a *doxastically justified belief* in that hypothesis, whereas the truth of the Boltzmann Brain hypothesis additionally makes it impossible for me to even have *propositional justification* to believe that hypothesis.[5] But, though I am convinced that this distinction between propositional and doxastic justification is an extremely important one in other contexts, I don't see any good reason to think that it is particularly useful here. First, it isn't completely clear what evidence I really *have* even in the hypoxia case (or in the hallucinatory drug case); when I'm hypoxic (or under the influence of the imagined hallucinatory drug), things seem perfectly normal to me, just as they do to Boltzmann Brains who are having experiences indistinguishable from those of ordinary observers.[6] Second, if I'm experiencing a particularly life-like dream, or if some conspiracy theory is true, I might not even *have* any evidence for those hypotheses, and yet I think that there are circumstances under which some such hypotheses shouldn't be simply disregarded (or assigned extremely low credences). And third, it seems clear to me that it is possible *in principle* for me to have some evidence in favor of the hypothesis that I am a Boltzmann Brain; for example, I could have the sorts of chaotic and disordered experiences that would presumably be quite typical of Boltzmann Brains and that are quite untypical of ordinary observers. I'll return to this point in Section 2.4 below.

None of this is to deny that the phenomenon of cognitive instability is an interesting or important one, or that it may often be epistemically relevant. But it seems to me that the cognitive instability of a particular hypothesis is not always a good reason to assign it a low credence. Rather, the cognitive instability of a hypothesis seems to be one way in which *a hypothesis, when true, is able to "hide itself" from rational discovery.*[7] If the hypotheses that I am dreaming or in The Matrix were true, those hypotheses would be very good at hiding themselves from rational discovery, as it would be difficult or impossible for me to acquire evidence that they are true. Similarly with the hypothesis that I am hypoxic: the perniciousness of hypoxia in the cases Elga refers to is precisely that, when a pilot is hypoxic, that fact often eludes rational discovery. And so too with conspiracy theories: by design, they are such that, when they are true, they are often difficult or impossible to get evidence for.

Different kinds and degrees of cognitive instability seem to correspond to varying senses in which, and extents to which, a hypothesis (when true) is able to elude discovery by making it (in varying ways and to varying extents) difficult or impossible for agents to form a reasonable belief that it is true. In some cases, this is because, when the hypothesis is true, the world is very much like (or very often like) the way the world is when the hypothesis is false; in such cases, it is very improbable for an event to occur to which even *could* count as a reason to believe or disbelieve the hypothesis. In other cases, the *entire* world might differ in important and widespread ways depending on whether the relevant hypothesis is true or false, and yet the world will tend to look the same *to observers* in either case. In still other cases, even though the truth of the hypothesis would impact the world in ways that are *in principle* detectable by observers, the truth of the hypothesis would also cause those very observers to fail to *notice* or *take appropriate account of* the ways in which the world is so impacted. These are all interesting – and interestingly different – cases of cognitive instability, but it strikes me as far too hasty to simply rule all of the relevant hypotheses out from serious consideration in one fell swoop.

Of course, many cognitively unstable hypotheses do indeed deserve low credences; indeed, many of them deserve *extraordinarily* low credences, and should be almost entirely disregarded in nearly every epistemic and practical context. Some conspiracy theories, for example, are so preposterously specific and ad hoc and reliant on coincidence that they cannot be taken seriously; even if they can be "cooked up" so that the likelihood that they assign to all of the available evidence is as high as one would like, their extraordinarily small prior probabilities guarantee that they should never be taken seriously in almost any practical or epistemic context other than the most extremely theoretical ones. In such cases, the cognitive instability of the relevant hypothesis is not *unrelated* to its low prior probability; it is entirely possible for specificity or ad hocness or reliance on coincidence, for example, to explain *both* the cognitive instability of a hypothesis *and* its low prior probability. But cognitive instability and low prior probability are logically distinct – even if oftentimes correlated – properties of a hypothesis. Cognitive instability, all by itself, is not a sufficient reason to reject a hypothesis. Fortunately, there are plenty of *other* good reasons to reject the hypothesis that I am a Boltzmann Brain; I will return to this issue in Section 2.4.

Finally, it is not even completely obvious that considerations of cognitive instability always rule out assigning high credence to cognitively unstable hypotheses. Take the hypothesis that all of my current experiences are being caused by electrical stimulations via the Matrix – the thought here was that this hypothesis is cognitively unstable because, if it's true, then any evidence that I have in its favor is unreliable. But suppose that I were to have the experience of a digital scroll moving across my visual field which reads "You are in The Matrix, and we're trying to get you out!!" If the Matrix Hypothesis is true, then that visual-scroll-evidence is Matrix-generated and hence in some sense an unreliable representation of what the world is like (since there is not in fact any text floating in front of me, as my visual experience represents there to be). But, it would be quite unreasonable to dismiss the Matrix Hypothesis, or to refuse to assign it a high credence, on this basis; even if the visual-scroll evidence is an unreliable representation of the world around me, it is still an excellent indication that I am in The Matrix, since it is hard to imagine what other hypothesis could explain the scroll. Similarly, if I were to have the sort of disordered or chaotic experiences that are (presumably) quite typical of Boltzmann Brains and quite atypical of ordinary observers, those experiences could (at least in principle) constitute strong evidence that I am a Boltzmann Brain; as I'll suggest in Section 2.4, the fact that I do *not* have these disordered and chaotic experiences should be seen as some evidence that I am *not* a Boltzmann Brain. So a great deal depends on the exact nature and strength of the putative evidence in favor of the hypothesis that we are Boltzmann Brains; unfortunately, I do not see any *general* reason to insist that nothing even in principle could count as evidence in favor of that hypothesis.

2.3 Hartle and Srednicki

Another reaction to the problems posed by Boltzmann Brains, defended by Hartle and Srednicki, is that "it is perfectly possible (and not necessarily unlikely) for us to live in a universe in which we are not typical" (2007, p. 123523-1). One important dialectical consequence of this claim is that, if it is true, then it would allow us to rationally hold both that the vast majority of the observers in the universe are Boltzmann Brains, and yet that *we* are (most likely) not; we would of course be quite atypical observers in this scenario, but on Hartle and Srednicki's view that is not (necessarily) a mark against the rationality of believing in such a scenario. According to Hartle and Srednicki, "[a] theory is not incorrect merely because it predicts that we are atypical" (ibid.). Thus, even if some cosmology were to entail that most of the observers in the universe are Boltzmann Brains, it would not follow that I am probably a Boltzmann Brain, and hence my rejection of the hypothesis that I am a Boltzmann Brain would not (necessarily) be a reason to reject that cosmology.

Hartle and Srednicki's argument relies to a significant extent on two analogies. First:

> Consider two theories of the development of planet-based intelligent life, based on the appropriate physics, chemistry, biology, and ecology. Theory A predicts that there are likely to be intelligent beings living in the atmosphere of Jupiter; theory B predicts that there are no such beings. Because Jupiter is much larger than the Earth, theory A predicts that there are today many more jovians than humans.
>
> Would we reject theory A solely because humans would not then be typical of intelligent beings in our solar system? Would we use this theory to predict that there are no jovians, because that is the only way we could be typical? Such a conclusion seems absurd.
>
> (2007, p. 123523-2)

Hartle and Sredinicki are correct that a presumption in favor of our own typicality would (all by itself) be no reason to reject the hypothesis that there are (likely to be) intelligent beings living in the atmosphere of Jupiter. But, I believe, this analogy profoundly misunderstands the nature and dialectical role of assumptions regarding our own typicality. The point is not that, *regardless of what evidence we have*, we should always take ourselves to be typical of any specified class of observers. If I have excellent evidence that I have just won the lottery, I certainly shouldn't reject the hypothesis that I've won simply because being a lottery winner would make me a quite atypical human being. And even if I'm certain that I just won the lottery, that doesn't give me any reason to believe that most of the other lottery players have probably won too (as that would make me a more typical lottery player), or that most other people are lottery players (as that would make me a more typical human in one sense), or that everyone in the world is now rich (as that would make me a more typical human in another sense), or any other such thing.

Rather, the point of the most plausible versions of epistemic presumptions in favor of typicality is that, if there are two observers at a world *who have the same evidence*, then I can't have any good reason to believe that I am one of them rather than the other. For example, Adam Elga's self-locating "Indifference Principle"[8] entails that if there are two "subjectively indistinguishable" agents at a world – i.e., individuals who "have the same apparent memories and are undergoing experiences that feel just the same" (2004, p. 387) – then a rational agent should be equally confident in the hypothesis that she is identical to the first of these agents as she is that she is identical to the second of these agents. As a consequence (at least in the finite case), if there are multiple subjectively indistinguishable observers at a world, then a rational agent should be more confident that she is one of the (more numerous) typical ones, rather than one of the (less numerous) atypical ones. A natural generalization of this principle to the infinite case will entail that if, among a set of subjectively indistinguishable observers, there is a subset of typical observers that has higher standard measure than the subset of atypical observers, then a rational agent should similarly be more confident that she is one of the typical observers. Indeed, a crucial component of several of the scientific and philosophical worries about Boltzmann Brains is precisely that (on certain cosmological assumptions) it is very likely that there are very many Boltzmann Brains *in precisely my current subjective state* in this world; if that is indeed the case, the thought goes, then my current subjective state would furnish me with no rational basis on which to conclude that I am an (atypical) ordinary observer, rather than a (typical) Boltzmann Brain.

However, Elga's Indifference Principle certainly does *not* entail that I should be more confident in hypotheses according to which there are fewer observers in the universe, or in hypotheses according to which there are fewer observers who are different from me in the universe.[9] If the humans and the jovians in the example would be in different subjective states, then Elga's Indifference Principle obviously doesn't apply. And even if they would be in the *same* subjective state, Elga's Indifference Principle still doesn't entail that we should have a low credence that there are jovians; all that it entails is that, on the assumption that there are jovians in the same subjective state as the humans, I should be more confident that I am one of the (more numerous) jovians than that I am one of the (less numerous) humans. So it seems to me that Hartle and Srednicki have an *extremely* different sort of typicality assumption in mind here from the ones that have currency in contemporary philosophical discussions.

Hartle and Srednicki's second analogy is as follows:

> Consider a model universe which has N cycles in time, $k = 1,...,N$. In each cycle the universe may have one of two global properties: red (*R*) or blue (*B*)...Two competing theories of this model universe are proposed. One, *all red* or *AR*, in which all the cycles are red, and another, *some red* or *SR*, in which some number of particular cycles are red and the rest are blue. We (an idealized observing system) seek to discriminate between these two theories on the basis of our data....Suppose that we (a particular observing system) observe red. Our data *D* is then (*E*,*R*), which in the context of the model could be more fully described as 'there is at least one cycle in which an observing system exists and the universe is red.
>
> (2007, p. 123523–4)

If there are a low number of cycles (and if the fraction or measure of red cycles according to *SR* is significantly lower than 1), then $p(E,R \mid SR)$ can be significantly lower than $p(E,R \mid AR)$, in which case (*E*,*R*) significantly favors *AR* over *SR* (at least on Hartle and Srednicki's Bayesian assumptions, which I'm happy to stipulate). But Hartle and Srednicki point out that, as the number of red cycles increases according to each hypothesis, it becomes overwhelmingly probable that there will be *at least one* red cycle regardless of whether *AR* or *SR* is true, even if the fraction or measure of red cycles is fairly small according to *SR*. Thus, as the number of red cycles increases, $p(E,R \mid SR) \approx p(E,R \mid AR)$, and hence (*E*,*R*) no longer significantly favors *AR* over *SR*. Hartle and Srednicki conclude that, on these assumptions, "[e]ven though the typical observing system in the *SR* theory is observing blue, our data provides no evidence that we are typical" (2007, p. 123523–5). Though this isn't made totally explicit, the analogical suggestion here is supposed to be that, as long as the conditions are such that the existence of *at least one* observer with my evidence is extremely likely, it is of no particular epistemological significance for a given cosmological theory that that theory entails that I am very atypical among all the observers who share my evidence. In particular, as long as a given cosmological theory entails that it is very likely that some ordinary observer with my evidence will exist at some point in the history of the universe, it is no mark against that theory if it *also* entails that the vast majority of observers in the universe with my evidence are Boltzmann Brains. If *SR* is true, most observers observe blue; but as long as it's very likely that someone will observe red, my observing red puts no rational pressure on me to reject *SR*. By analogy, if a particular cosmological theory is true, most observers are Boltzmann Brains; but as long as it's very likely that there will be an ordinary observer, my confident belief that I'm an ordinary observer puts no rational pressure on me to reject the cosmological theory.

However, it seems to me that Hartle and Srednicki's argument again contains a critical error, which can be traced back to the assumption that the relevant data is the proposition "there is at least one cycle in which an observing system exists and the universe is red." This is a flagrant violation of the Principle of Total Evidence,[10] according to which we should always take account of the *strongest* statement of our evidence that we have available to us. And the *strongest* statement of our evidence isn't the claim that there is *at least one* cycle in which an observing system exists and the universe is red; it is that *this* observing system – i.e., me – exists and observes red. And *that* is much more likely if *AR* is true than if *SR* is true, and hence (contra Hartle and Srednicki) our evidence confirms *AR* over *SR*.[11]

For comparison, suppose that 100 conscious observers are about to be created, and that either (*AR*) all 100 of them are going to be put in red rooms, or that (*SR*) only one of them of them is going to be put in a red room and the remaining 99 are going to be put into blue rooms. I wake up in a red room and am told all of this. Can there be any serious doubt that my observing red is strong evidence for *AR* over *SR*? After all, if *AR* is true, then I will certainly observe red; and if *SR* is true, then I am rather unlikely to observe red. It is irrelevant that, on either theory, it is certain that *at least one observer* will observe red. That's a weaker statement of my evidence than what I know, and I would violate the Principle of Total Evidence by conditionalizing on it and nothing stronger.

2.4 The Badness of Boltzmann Brains

Why *should* I reject the hypothesis that I am a Boltzmann Brain, and what implications does this have for cosmology? There are two good (if fairly obvious) reasons to reject the hypothesis that I am a Boltzmann Brain. First, the physical probability is staggeringly low that I would fluctuate into existence. And second (to return to a point from Section 2.2), even if I did fluctuate into existence, it is overwhelmingly improbable that I would be having anything like the ordered and coherent stream of thoughts and experiences that I am in fact having; the vast majority of conscious Boltzmann Brains would have wildly disordered and incoherent thoughts and experiences. The relevance of these points to different cosmologies should be analyzed separately.

Sometimes, cosmological consequences related to Boltzmann Brains are appealed to in order to argue that the universe must not be infinite in time or space or both.[12] If the universe is infinite, it looks to follow that there will be infinitely many Boltzmann Brains. Suppose, in addition, that we are considering a cosmology according to which the set of all Boltzmann Brains vastly outnumbers (or: has a vastly higher asymptotic density[13] than) the set of ordinary observers. And suppose even further we are considering a cosmology according to which the set of all Boltzmann Brains *with ordered and coherent streams of thoughts and experiences* (or perhaps even stronger: with *my* total evidence) vastly outnumbers (or has a vastly higher asymptotic density than) the set of all ordinary observers with ordered and coherent streams of thoughts and experiences.

If I were certain (or nearly certain) that such a cosmology were true, then it seems unavoidable (for reasons related to the discussion of typicality in Section 2.3) that I should be overwhelmingly confident that I am one of the Boltzmann Brains with ordered and coherent streams of thoughts and experiences, rather than an ordinary observer. But, notwithstanding all of that, the fact remains that, if such a cosmology were true, it would be overwhelmingly probable (again for reasons related to the typicality considerations from Section 2.3) that I would have had the *disordered* and *incoherent* streams of thoughts and experiences that the vast majority of Boltzmann Brains (and hence the vast majority of observers in the universe) have. And since my *actual* thoughts and experiences are so highly ordered and coherent, I take myself to have strong evidence against a cosmology like that – evidence that doesn't depend on the *premise* that I'm not a Boltzmann Brain. The point here is not that there couldn't be a Boltzmann Brain that has ordered and coherent thoughts and experiences; indeed, if the universe were infinite and constantly fluctuating, there would (with probability 1) be *infinitely many* such Boltzmann Brains. And the point is not that it is more improbable for a particular Boltzmann Brain to have ordered experiences than it is for that Boltzmann Brain to come into existence to begin with. The point, rather, is that on the assumption that *I* am a Boltzmann Brain, it is incredibly unlikely that *I* would have such ordered and coherent thoughts and experiences, regardless of how likely it is that I would come into existence as a Boltzmann Brain to begin with. And that gives us reason to prefer a cosmology on which it's more probable that a randomly selected observer would have ordered thoughts and experiences.

Considerations having to do with Boltzmann Brains are *also* sometimes taken to cause problems for cosmologies according to which the state of our entire universe, or the portion of the universe in which we live, is the result of a quantum or thermodynamic fluctuation.[14] As I understand this concern, it is independent of whether the entire universe is infinite or not; whereas the prior concern was specifically about cosmologies according to which the universe is infinite, this set of worries applies even to cosmologies according to which the universe is finite.

The idea here is that, though the fluctuation of a Boltzmann Brain is indeed wildly improbable, the fluctuation of an *entire universe* or even of the *portion* of the universe we're able to observe into a low-entropy state is even *more* wildly improbable, and by a large degree, since low-entropy states of universes (or large portions thereof) require larger fluctuations in order to arise than Boltzmann Brains do. Thus, it is overwhelmingly more probable for a Boltzmann Brain with my current experiences to fluctuate into existence than for an entire universe (or a large portion thereof) to fluctuate into a low-entropy state, and hence (if I accept the cosmology under consideration) I should be more confident in the former hypothesis than in the latter one.

This reasoning strikes me as overwhelmingly compelling. Moreover, I am completely persuaded by Albert's (2000) and Carroll's (2010, chapters 8–9) arguments that we should accept the "Past Hypothesis" that the observable universe began in a state of very low entropy. Though a full discussion of the Past Hypothesis is not possible here, one of the central virtues of the Past Hypothesis is that it explains why, e.g., a photograph (or memory) is very likely to have been caused by the actual events that it represents; even though the most likely way *in the space of all possible evolutions of the universe* for the photograph to have come into existence is for it to have randomly fluctuated from a higher-entropy past, it is also the case that the most likely way *in the space of all evolutions of the universe from a low-entropy beginning* for this photograph to have come into existence is for it to have been caused by the event it represents. Similarly for me: even though the most likely way in the space of all possible evolutions of the universe for me to have come into existence is to have randomly fluctuated from a higher-entropy past, it is also the case that the most likely way in the space of all evolutions of the universe from a low-entropy beginning for me to have come into existence is through a process characteristic of ordinary observers. So, reasons to accept a cosmology that includes the Past Hypothesis (of which I think there are powerful ones) are also reasons to reject the hypothesis that I am a Boltzmann Brain.

The question remains, of course, of why the Past Hypothesis is true. And I think that one of the lessons here is that it will not do to say that the Past Hypothesis was itself true as a result of random fluctuation; if *that* were the only way that the Past Hypothesis could have been true, then I should prefer the hypothesis that I am a Boltzmann Brain on the grounds that this latter hypothesis is so much more probable. But it seems to me that the Inflation Theory[15] – according to which there was a period of exponential expansion of the universe during its first few moments – offers some realistic hope of explaining the truth of the Past Hypothesis without invoking the sorts of minuscule probabilities that would be associated with the Past Hypothesis being true as a result of random fluctuation. In brief, the idea here is that, prior to Inflation, the portion of the universe that was to become the observable universe was microscopic, and that quantum (and perhaps thermal) fluctuations on this microscopic scale expanded during Inflation to regions of low entropy that would make the Past Hypothesis true. The Inflation Theory has had many successes,[16] and I think that there are grounds for a great deal of optimism about both its truth and its capacity to explain our manifest experience without appealing to any wildy improbable fluctuations.

Notes

1 Some cosmologists have argued that Boltzmann Brains that arise as quantum fluctuations in the vacuum pose more serious problems than those that arise as thermal fluctuations – see, e.g., Davenport and Olum (2010). In this paper, I will ignore any differences that exist between these different sorts of Boltzmann Brains.

2 For a non-standard view of quantum fluctuations in de Sitter space, see Boddy, Carroll, and Pollack (2017); they argue that quantum fluctuations in isolated quantum systems are an epistemic phenomenon rather than a genuine physical one, and hence that Boltzmann Brains won't appear in the true de Sitter vacuum. However, as far as I know, there is absolutely no reason to doubt the physical possibility of Boltzmann Brains that arise by way of thermal fluctuation out of a thermal equilibrium.

3 Or, if such a thing is possible, to assign *no particular credence at all* – the credal equivalent of "withholding judgment" – to the hypothesis.

4 Thanks to Ram Neta for helpful discussion of this proposal.

5 See Ichikawa and Steup (2014, Section 1.3.2).

6 See Neta (2008) for a discussion what it means for an agent to "have" some particular piece of evidence.

7 I do not claim that cognitive instability *just is* the ability of a hypothesis to hide itself from rational discovery, nor do I claim that the *only* way for a hypothesis to hide itself from rational discovery is to be cognitively unstable. A hypothesis might be able to hide itself from rational discovery, for instance, simply by having a low prior probability and by making all of the same empirical predictions as a hypothesis with a high prior probability; no cognitive instability would be required. I claim only that cognitive instability is one way for a true hypothesis to evade rational discovery. This ability may well make it unreasonable for observers to assign the hypothesis a very high credence, but I am arguing here that it does *not* automatically make it reasonable for observers to assign the hypothesis a very low credence.

8 See Elga (2000, 2004).

9 It is controversial whether some version of the "Self-Indication Assumption" formulated in Bostrom (2002) is true: "Given the fact that you exist, you should (other things equal) favor hypotheses according to which many observers exist over hypotheses on which few observers exist." (p. 66) I think that the most plausible version of the SIA adds a clause about sharing your evidence: "Given the fact that you exist, you should (other things equal) favor hypotheses according to which many observers who have your evidence exist over hypotheses on which few observers exist who have your evidence." If this latter principle is true, and if the humans and jovians in Hartle and Srednicki's analogy share the same evidence, then there is a case to be made that our existence provides some reason to believe that there *are* jovians.

10 See Carnap (1947).

11 For arguments along similar lines (though applied in somewhat different contexts), see Kotzen (2013) and White (2000).

12 See, e.g., Page (2007, 2008a, 2008b, 2008c). For useful discussion, see also Dyson, Kleban, and Susskind (2002); Bousso and Freigovel (2007); Linde (2007); Vilenkin (2007); and Banks (2007).

13 On the assumption that the set of observers in the universe is countably infinite, we cannot appeal to the standard Lebesgue measure here, since the Lebesgue measure can be definted only in spaces that can be represented as Euclidean n-dimensional *real-valued* spaces, and if the set of observers in the universe is countable, then the space of possible numbers of observers is natural-number-valued. Thus, we must appeal

to asymptotic densities, which yield the "proportion" of natural numbers up to *n* that have some property, in the limit as *n* approaches ∞. See Nathanson (2000) and Tenenbaum (1995) for discussions of asymptotic densities. See Buck (1946) for a discussion of the analogy between measures and asymptotic densities.

14   See, e.g., Albrecht and Sorbo (2004), and Carroll (2016, Chapter 11).

15   The Inflation Theory was developed by Alan Guth, Andrei Linde, Paul Steinhardt, and Andreas Albrecht.

16   For instance, the Inflation Theory is often thought to explain the nearly-flat geometry of the universe, the uniformity of the cosmic background radiation, and the absence of stable magnetic monopoles. For an accessible overview, see NASA (n.d.).

References

Albert, David Z. (2003). *Time and chance*. Cambridge, MA: Harvard University Press.

Albrecht, Andreas, & Sorbo, Lorenzo. (2004). Can the universe afford inflation? *Physical Review D, 70*(6). [arXiv:hep-th/0405270]

Banks, Tom. (2007). Entropy and initial conditions in cosmology. [ArXiv preprint at arXiv:hep-th/0701146]

Boddy, Kimberly K., Carroll, Sean M., & Pollack, Jason. (2017). Why Boltzmann brains don't fluctuate into existence from the De Sitter vacuum. In Khalil Chamcham, Joseph Silk, John D. Barrow, et al. (Eds.), *The philosophy of cosmology* (pp. 228–240). Cambridge, UK: Cambridge University Press. [arXiv:1505.02780 [hep-th]

Bostrom, Nick. (2002). *Anthropic bias: Observation selection effects in science and philosophy*. New York: Routledge.

Bostrom, Nick. (2003). Are we living in a computer simulation? *The Philosophical Quarterly, 53*(211), 243–255.

Bousso, Raphael, & Freivogel, Ben. (2007). A paradox in the global description of the multiverse. *Journal of High Energy Physics, 06*, 018. [arXiv:hep-th/0610132]

Buck, R. Creighton. (1946). The measure theoretic approach to density. *American Journal of Mathematics, 68*(4), 560–580.

Carnap, Rudolf. (1947). On the application of inductive logic. *Philosophy and Phenomenological Research, 8*(1), 133–148.

Carroll, Sean M. (2010). *From eternity to here: the quest for the ultimate theory of time*. New York, NY: Penguin.

Carroll, Sean M. (2016). *The big picture: on the origins of life, meaning, and the universe itself*. Boston, MA: Dutton.

Davenport, Matthew, & Olum, Ken D. (2010). Are there Boltzmann brains in the vacuum? [ArXiv preprint at arXiv:1008.0808 [hep-th]]

Descartes, Rene. (n.d.). *Meditationes de prima philosophia, in qua Dei existentia et animae immortalitas demonstrantur [Meditations on first philosophy]*. Paris: Michel Soly.

Dyson, Lisa, Kleban, Matthew, & Susskind, Leonard. (2002). Disturbing implications of a cosmological constant. *Journal of High Energy Physics, 10*, 011. [arXiv:hep-th/0208013]

Elga, Adam. (2000). Self-locating belief and the Sleeping Beauty problem. *Analysis, 60*(266), 143–147.

Elga, Adam. (2008). *Lucky to be rational*. Unpublished manuscript. Available at www.princeton.edu/~adame/papers/ bellingham-lucky.pdf

Elga, Adam. (2004). Defeating Dr. Evil with self-locating belief. *Philosophy and Phenomenological Research, 69*(2), 383–396.

Hartle, James B., & Srednicki, Mark. (2007). Are we typical? *Physical Review D, 75*(12), 123523. [arXiv:0704.2630 [hep-th]]

Ichikawa, Jonathan J., & Steup, Matthias. (2014). The analysis of knowledge. In Edward N. Zalta (Ed.), *The Stanford Encyclopedia of Philosophy (Spring 2014 Edition). Available at* http://plato.stanford.edu/archives/ spr2014/entries/knowledge-analysis/.

Kotzen, Matthew. (2013). Multiple studies and evidential defeat. *Noûs, 47*(1), 154–180.

Kotzen, Matthew. (2012). Selection biases in likelihood arguments. *The British Journal for the Philosophy of Science, 63*(4), 825–839.

Linde, Andrei D. (2007). Sinks in the landscape and the invasion of Boltzmann Brains. *Journal of Cosmology and Astroparticle Physics, 01*, 022. [arXiv:hep-th/0611043]

Nathanson, Melvyn B. (2000). *Elementary methods in number theory*. New York: Springer-Verlag.

National Aeronautics and Space Administration [NASA]. (n.d.). *Our universe: what is the inflation theory?* Available at http://wmap.gsfc.nasa.gov/universe/bb_cosmo_infl.html

Neta, Ram. (2008). What evidence do you have? *The British Journal for the Philosophy of Science, 59*(1), 89–119.

Page, Don N. (2007). Susskind's challenge to the Hartle–Hawking no-boundary proposal and possible resolutions. *Journal of Cosmology and Astroparticle Physics, 01*, 004. [arXiv:hep-th/0610199]

Page, Don N. (2008a). Is our universe decaying at an astronomical rate? *Physics Letters B, 669*(3–4), 197–200. [arXiv:hep-th/0612137]

Page, Don N. (2008b). Is our universe likely to decay within 20 billion years? *Physical Review D, 78*(6), 063535. [arXiv:hep-th/0610079]

Page, Don N. (2008c). Return of the Boltzmann brains. *Physical Review D, 78*(6), 063536. [arXiv:hep-th/0611158]

Tenenbaum, Gérald. (1995). *Introduction to analytic and probabilistic number theory. Cambridge Studies in Advanced Mathematics*. Cambridge: Cambridge University Press.

Vilenkin, Alexander. (2007). Freak observers and the measure of the multiverse. *Journal of High Energy Physics, 2007*(01), 092. [arXiv:hep-th/0611271]

White, Roger. (2000). Fine-tuning and multiple universes. *Noûs, 34*(2), 260–276.

Study Questions for Part I

1. What is a Boltzmann brain? In what are they similar and in what are they different from ordinary observers?
2. What is the standard argument against cosmologies dominated by Boltzmann brains? Where does this argument go wrong according to Carroll?
3. What is "cognitive instability"? Why does it pose a problem for cosmologies dominated by Boltzmann brains according to Carroll? Why, according to Kotzen, is it not a sufficient reason to reject a hypothesis?
4. Hartle and Srednicki argue that "[a] theory is not incorrect merely because it predicts that we are atypical." Explain one of the examples they use to support this claim and how it is supposed to bear on the Boltzmann brains problem. Does Kotzen think the analogy works? Why?
5. How can issues related to Boltzmann brains be used to argue against cosmologies in which time, space, or both are infinite?

Part II

Does Mathematical Explanation Require Mathematical Truth?

3 Mathematical Explanation Requires Mathematical Truth

Christopher Pincock

3.1 An Explanatory Argument for Mathematical Platonism

One way to appreciate the challenge that mathematical explanation poses for the philosophy of science is to highlight a contrast between two sorts of naturalism. Quine advocated a thoroughgoing naturalism that counsels that philosophers should only criticize science using the tools and standards found within science. Quinean naturalism thus rejects any foundationalist first philosophy that would presume to dictate what knowledge is or how science should be done. In his book *Word and Object* Quine deploys a famous simile from Neurath as one of his epigraphs to drive this point home: "We are like sailors who must rebuild their ship on the open sea, never able to dismantle it in dry-dock and to reconstruct it there out of the best materials" (Quine, 1960). Neurath anticipated Quine's naturalism in a number of respects, but they ultimately clashed on the issue of metaphysics. For Neurath followed his simile with the claim that "Only the metaphysical elements can be allowed to vanish without trace" (Neurath, 1934/1966, p. 201). In a properly naturalized science, Neurath supposed, metaphysical claims would be removed because they contributed nothing to the success of science.

Quine disagreed with Neurath on the status of metaphysics. For Quine, metaphysical commitments do earn their keep through their contribution to the overall success of science. This point is explicit in "Two Dogmas of Empiricism": "Ontological questions … are on a par with questions of natural science" (Quine, 2004, pp. 52–53). Whether or not we should believe in neutrinos is settled by determining which theory best accommodates our interactions with the world. Similarly, for Quine, questions about the existence of platonic mathematical entities like numbers and sets should be answered by considering our best scientific theory. If this theory makes use of statements that imply the existence of these entities, then we should believe that they exist. Metaphysical elements may turn out to be central to the success of science, and the traditional metaphysical questions that Neurath had sought to expunge are revived in a new form.

Quine's argument for the existence of platonic mathematical entities can be resolved into two parts. First, there is Quine's criterion for ontological commitment. Quine claims that a scientific theory is a collection of statements, but that these statements must first be regimented in a definite way if one is to assess what their truth requires of the world. For Quine this regimentation uses only first-order logic. Once this process is complete, there is a determinate test that decides if that theory posits the existence of entities of kind F. One asks if the statements of the theory entail "There are Fs" (or in symbols: $(\exists x)$ (Fx)).

Regimentation of scientific activity leads to a collection of conflicting theories that a scientist or philosopher could assent to. The second part of Quine's argument considers the various theoretical virtues that we wish a theory to have. These virtues are determined by an examination of how science is done. Quine's naturalism means that it is only scientific practice that can settle which theory is the best. At one point he proposed five virtues: conservatism, generality, simplicity, refutability and modesty (Quine & Ullian, 1978). We aim for general theories that make sense of many phenomena using the simplest viable means. But the generality of the theory should not sacrifice the testability and refutability of its central claims in light of future experiments. Similarly, our theories should be modest in that they should not make unwarranted leaps into speculation. Quine also emphasized, under the heading of "conservatism," the value of retaining the beliefs that one begins with.

Quine's considered position, then, is that the regimented theory that maximizes the theoretical virtues will posit the existence of platonic mathematical entities. Arguments of this sort have come to be known as indispensability arguments for mathematical platonism. The main question for these arguments is whether it would be better to opt for a revised theory that limited its commitments to platonic entities, or perhaps eliminated them entirely. This sort of revised theory might be less simple, less general and less conservative than the platonic theory that Quine preferred. But it would seem to score well on the virtues of modesty and refutability. Platonic entities are a seemingly strange addition to our metaphysics, and their features elude any direct experimental testing.

The upshot of this debate is that there is no simple recipe for arriving at the best theory. Quinean platonists embrace a scientific argument for the existence of mathematical entities that relies on a certain list of theoretical virtues as well as a criterion for ontological commitment. Nominalists, who deny the existence of any abstract entities, thus have two points of attack. They may revise or deny Quine's test for the ontological commitments of a scientific theory. Or the nominalist may question the application of the Quinean theoretical virtues. The best theory, all things considered, may not require any abstract objects.[1]

This standoff between Quinean platonists and nominalists has motivated a new debate about inference to the best explanation (IBE). Many scientists take the explanatory power of a proposed theory to be part of the evidence that the theory is true. The Big Bang theory is said to provide the best explanation for the observed pattern of background radiation. The theory that a bolide (comet or asteroid) hit the Earth near Mexico is presented as the best explanation for the extinction of the dinosaurs, which is in turn accepted as the best explanation for the observed distribution of dinosaur fossils. An explanatory indispensability argument presents these explanatory considerations as a sufficient reason to accept the existence of platonic mathematical entities. For it appears that many scientific theories make essential use of mathematical entities when they explain observed phenomena. These explanations are the best available because they exhibit various explanatory virtues. These virtues include the generality and simplicity that we have seen already in connection with theories. But now with explanations we want an explanation to explain as much as possible using the simplest means.

The advocate of the explanatory indispensability argument points to specific explanations that rely on mathematical entities. One example that has received extensive discussion is the geometric structure of the honeycombs that bees construct to store their honey. A two-dimensional cross-section of these honeycombs shows a hexagonal tiling pattern, and scientists seek an explanation of this trait. The mathematical explanation is that this trait is an evolutionary adaptation of the bees that allows them to cover the most area using the minimum amount of wax. This is the fittest trait because it does the most with a costly resource. And this explanation is said to be the best explanation because it exhibits virtues like generality and simplicity.

Varro's *On Agriculture* from 36 BC presents this explanation as superior to one that cites the number of legs that bees have: "Does not the chamber in the comb have six angles, the same number as the bee has feet? The geometricians prove that this hexagon inscribed in a circular figure encloses the greatest amount of space" (Cato & Varro, 1934, p. 501, noted by Hales, 2000, p. 448). The mathematical claim at the heart of the explanation is that hexagons maximize the ratio between the area covered and the length of the edges used in the tiling. A proof of this theorem for the special case of regular polygons appears in Pappus' *Mathematical Collection* from around 340 AD, but was probably known much earlier (Nahin, 2007, appendix C). The theorem was proved for all polygons only in 1999 by Thomas Hales. When regimented, it entails the existence of mathematical entities such as hexagons and ratios. The defender of the explanatory indispensability argument concludes that our best science provides adequate support for the existence of mathematical entities. Scientists wish to explain observed phenomena, and to do so they must accept the existence of the Big Bang, a bolide hitting the Earth, hexagons and ratios. There is thus supposed to be no principled way to be a defender of IBE and also a nominalist.[2]

3.2 Explanation and Truth

IBE takes the explanatory power of a proposed explanation as evidence that the proposed explanation is in fact true. The Quinean platonist may be content to motivate this connection through an appeal to scientific practice: scientists, we are told, do use explanatory power as part of their evidence. If so, any naturalist should take this form of inference as valid. This argument ignores the critical attitude that is implicit in naturalism. For the naturalist can and should reject a given instance of IBE if it clashes with our scientific methods and standards. Many scientists seem uncomfortable with an unrestricted form of IBE that takes being the best available explanation as the only factor in theory choice. An example that has been emphasized by Maddy concerns the existence of atoms (Maddy, 1998). In the first half of the nineteenth century Dalton proposed the existence of atoms as part of his explanation for the proportions found in various chemical reactions. However, this attempted IBE argument for the existence of atoms was unconvincing for most scientists. It was only at the beginning of the twentieth century, with the work of Perrin in particular, that a persuasive IBE argument was found. This suggests that a proposed explanation must meet additional restrictions beyond its potential explanatory power before it is accepted by the scientific community. Once we consider what sort of restrictions should be in place, then it becomes reasonable to worry that a novel ontological commitment is excessive.[3]

The major step in any IBE is the consideration of how the correctness of the proposed explanation would account for the phenomenon in question. In the extinction of the dinosaurs case, it must be clear how a bolide hitting the Earth at that place and time would make sense of the later extinction of the dinosaurs. A similar point holds for the honeycomb case. If we consider the possibility that a hexagon really is the best basic shape with which to build a honeycomb, then this should illuminate the presence of that shape in the bee's honeycombs. To say that these proposed explanations are the best is to say that, of all the available proposals, these ones exhibit the explanatory virtues to the highest degree. These virtues do not entail that the best explanation is true. For all we can tell, the correct explanation of these phenomena might not have been formulated for our consideration. Another possibility is that the correct explanation has been formulated, and yet as presented it fails to show the explanatory virtues. The defender of IBE can admit the fallibility of their inference. However, there must be something special about uses of IBE that lead us astray. Otherwise, it would be unscientific to persist in using this form of ampliative inference.

Like any other kind of evidence, the explanatory evidence that drives these inferences can be degraded. Two sorts of scenarios are worth considering. In the first scenario, we receive independent evidence that a central claim of the best explanation is in fact false. In the second kind of situation, the original best explanation is shown to be inadequate through the formulation of an even better explanation. In both cases, a defender of IBE should advise a shift away from the original claims.

To see how such cases would work, consider again the bolide explanation for the extinction of the dinosaurs. Suppose that the only evidence for the existence of this bolide impact is the way that claims about the bolide contributed to the best explanation of the extinction. Scientists might arrive at independent evidence that indicates that no bolide could have struck the Earth during the relevant time period. If this evidence was strong enough, then it would be prudent to withdraw our commitment to that bolide, even if it left us with no explanation of the extinction of the dinosaurs. In a second kind of case, suppose that additional investigations led to a better explanation of the extinction in terms of volcanic activity rather than a bolide. This might involve a more detailed understanding of what drives volcanic activity along with new indications of a rise in volcanic activity right before the dinosaurs went extinct. Again, as in the first case, the rational thing for scientists to do would be to withdraw their belief in the bolide, and opt for the volcano explanation.

The honeycomb explanation has a different character than the bolide explanation, but the same points apply concerning the defeasibility of explanatory evidence. For the first kind of case, we need only suppose that mathematicians find a flaw in Hales' recent proof or even the ancient proof presented by Pappus. This could lead to a new theorem that some other polygon maximized the ratio between its area and the length of its perimeter. If the support for this new theorem was strong enough, then the original explanation of the hexagonal shape of the bees' honeycomb would have to be rejected. Also, if the only support for the existence of hexagons and ratios was this explanatory evidence, then the prudent scientist would withdraw belief in those platonic entities. A slightly different result obtains in the other kind of scenario. Suppose an innovative scientist proposed a new explanation of the hexagonal shape of the honeycomb cells that tied that shape to the fact that bees have six legs. For us it is a mere coincidence that the cells have six sides and the bees have six legs. So this proposal scores very poorly as a proposed explanation as there is no illuminating connection between these numbers. But some future scientific development could fill in this connection, and come to surpass the current geometric explanation in explanatory power. If this happened, then scientists would not have a reason to accept the hexagons and ratios of the original explanation. They would instead opt for a commitment to the numbers that are central to the new explanation. That is, they would shift from a belief in one sort of platonic entity to another sort of platonic entity.[4]

Thinking through these sorts of scenarios highlights the strong link between genuine explanation and truth. IBE is a form of ampliative inference that, like all inference, aims at truth. To properly carry out the inference, we do not need to know that the conclusion we are drawing is in fact true. But the inference should be corrected and reconsidered as new evidence comes in. If we have accepted claim C due to its presence in the best available explanation, then we should reconsider C if it turns out that C is not part of the best available explanation. In one scenario, we receive independent evidence that C is false. If this new evidence is strong enough, then we should abandon our commitment to C and also to the explanation that we previously accepted. This may leave us with no explanation of the phenomenon in question. In a second scenario a better explanation is proposed and this new explanation turns out to lack C. In this case, we should switch to this better explanation and drop our acceptance of C.

This pattern of acceptance and rejection only makes sense if we suppose that genuine explanations are composed of true claims. For the special case of mathematical explanations of physical phenomena, mathematical explanation requires mathematical truth. It turns out that there are accounts of scientific explanation that do not require truth. For example, van Fraassen's pragmatic theory of explanation insists that explanations are answers to why-questions (van Fraassen, 1980). Van Fraassen claims that we can use the theories that we accept to answer these why-questions even if we do not believe that these theories are true. Given the discussion of this section, it is unsurprising to find that van Fraassen rejects IBE (van Fraassen, 1989). On his approach, there is no link between a genuine explanation and truth. And so there is no way to motivate IBE.

A defender of IBE should maintain that explanation requires truth. Many scientific realists go further than this. They suppose that IBE can be used to characterize the natures of the entities involved in their best explanations. However, in scientific uses of IBE, there is widespread caution in using explanatory considerations to settle contested questions. In the bolide case, even though the bolide hypothesis was accepted based on explanatory considerations, the nature of the bolide that hit the Earth remained subject to debate. As early as 1980, the presence of iridium in the geological record at the time of the dinosaur extinction was sufficient to convince many scientists that an impact of some sort caused the extinction. This explanation was judged to be better than its primary competitor, the volcanic activity hypothesis. It was only in 1991, when the Chicxulub crater was identified, that the size and composition of the bolide was ascertained with sufficient confidence. And it was only this additional information that narrowed down the nature of the bolide to an asteroid of a specific size (10 km wide). Debates continue concerning the significance of the timing of the asteroid's impact. While some argue that the impact would have caused the extinctions no matter when it occurred, others maintain that an earlier impact would not have led to the extinction of the dinosaurs (Schulte et al., 2010; Brusatte 2015; Brusatte et al., 2015).

We can make sense of the change from the initial acceptance of the truth of some claim to the later acceptance of the existence of a new type of entity if we suppose that IBE is essentially eliminative. On this approach, when a scientist accepts a hypothesis using IBE they are using explanatory considerations to eliminate competitors. For example, in 1980 the iridium was used to eliminate the volcano hypothesis in favor of the impact hypothesis. But this choice in favor of the impact hypothesis did not yet settle the character of the impact, and so scientists remained agnostic on these additional details. It was only when a specific crater was identified in 1991 that a new IBE argument was possible that pinned down the character of the body that impacted the Earth. This additional information could only be explained by a certain kind of asteroid, and so the hypotheses concerning other sorts of bolides were eliminated from consideration.

These reflections on IBE raise some problems for the Quinean platonist. Even if one accepts the explanation-truth link, one needs independent motivation to draw any conclusions about the natures of the mathematical entities from their putative role in these scientific explanations. The problem is that the mathematical entities figure in these explanations solely in terms of their abstract structure. In the bees case, the theorem requires only that a certain ratio hold between areas and perimeters. It does not require anything of the geometric entities themselves. Similarly, a mathematical explanation that makes use of arithmetic will turn only on the structure of the natural numbers and their structural relationships. This will not pin down the character of the natural numbers. The explanation has its virtues, and the mathematical claims that figure in the explanation may be true, no matter what additional intrinsic features the mathematical entities may possess. Unlike the asteroid case, there is no forthcoming information that would allow scientists to eliminate some of the hypotheses concerning the natures of these entities, and so support a single remaining existential hypothesis. I conclude that it is illegitimate for the advocate of the explanatory indispensability argument to use their argument to say anything more than that some mathematical claims are true.[5]

3.3 Truth and Fiction

The main alternative to the claim that mathematical explanation requires mathematical truth is known as fictionalism. Fictionalists propose a link between the mathematical claims that we accept and fiction. Consider, for example, Pappus' theorem that hexagons are the regular polygons that cover a given area using the shortest perimeter. A traditional platonist insists that this is a true claim about geometric entities and cites the proof as conclusive evidence for this truth. A Quinean platonist worries that the axioms used to obtain the proof may be false, and so rejects traditional platonism. But the Quinean platonist would accept the theorem (and perhaps also the axioms) if that package of claims made important contributions to scientific explanation. A fictionalist remains unconvinced. On their approach, false claims may be used in science, even in explanations, and even when we think they are false. This is because these claims appear in a coherent collection of fictional claims. Fictional claims are true in a given fiction, but may be false of the real world or "outside" the fiction. The thought is that fictional claims can explain features of the real world. Their falsity is no barrier to their explanatory power.

No matter how the collection of fictional claims is specified, the fictionalist faces the challenge of clarifying how claims that are false can figure in a genuine explanation. One version of this challenge has been pressed by Colyvan in his discussion of "easy road" nominalism. An easy road nominalist uses mathematics in scientific explanations while denying that these mathematical claims are true. This differs from a "hard road" nominalism that tries to formulate nonmathematical versions of the scientific explanations that we accept. The hard road nominalist can offer their reformulations for comparison with the original mathematical explanations and presumably try to make the case that the nominalist explanations are superior. The easy road nominalist is not able to do this because they work with the original explanations. What they add is some kind of qualified acceptance of the truth of the claims that appear in these explanations. The purely mathematical claims are relegated to the purely fictional, and so are said to explain even though they are false.

Colyvan's challenge is quite simple: "when some piece of language is delivering an explanation, either that piece of language must be interpreted literally or the non-literal reading of the language in question stands proxy for the real explanation" (Colyvan, 2010, 300). The fictionalist proposes a fictional or nonliteral interpretation of the claims found in the explanation and appears to deny that there is any additional "real" explanation backing it up. However, if this is all the fictionalist says, then they are not able to vindicate some of the core features of an explanation. In the dinosaurs case, for example, the asteroid proposal explains the extinction by linking the asteroid event with the later extinction. If we accept the explanation, then we take ourselves to have a reason for the extinction: the extinction occurred because that asteroid hit the Earth. It is not an explanation to be told that an asteroid hit the Earth, but that this should not be taken literally as it is only part of a collection of fictional claims. Colyvan's challenge, then, is to supplement this bare proposal with some more substantial account of how fictional claims may explain.

One suggestion has been made by Elgin (2009). She notes that fictional entities may exemplify properties that are also instantiated in the real world. So we may compare a purely fictional scenario with an actual one, and provide the explanation through this sort of comparison. This proposal seems to work best with fictional characters such as Sherlock Holmes. Holmes' superior intelligence generates various traits such as impatience and an inability to show affection for his close friend Watson. The fiction thus illustrates a link between certain kinds of character traits. We see how intelligence can bring about impatience with others. When we see these features instantiated in the real world in our colleagues, then the fiction can be useful in explaining this connection. This colleague is impatient with us because she is so intelligent, just as with Holmes.

I maintain that this is not an adequate response to Colyvan's challenge. To see why, we can distinguish how one learns an explanation from the explanation itself. In the case just reviewed, it would be wrong to say that my colleague is impatient because Holmes is impatient, or that my colleague is impatient because Holmes is impatient due to his intelligence. My colleague is impatient because she is so intelligent. That is the explanation. I may learn the explanation by reading the literary fiction in the sense that I become aware of a potential explanatory link between these traits by these means. But the explanation itself makes no use of the fiction. If there is a link between these traits, then my colleague's impatience is due to her intelligence. This is a true explanatory claim.

A different proposal has been defended by Leng (2010, 2012). Leng's view can be seen as an attempt to extend the point made at the end of Section 3.2 about the limited role of the natures of mathematical entities in mathematical explanations. If we suppose that these explanations require only an abstract structural relation between mathematics and the real world, then it might seem like this relation could obtain between the imagined subjects of fictional claims and the real world. On this view, one side of this relation does not exist and the claims we make about it are all false. Nevertheless, it is in virtue of this relation that a real world fact is explained.

I claim that Leng's structural proposal is also not an adequate response to Colyvan's challenge. Consider again the honeycomb case. The explanation claims that the fittest trait for the bees is to construct their honeycombs in hexagons. Part of the explanation makes a link between these shapes and the minimal use of wax. On the platonist reading, the mathematical theorem is true and it is the reason that this trait is the fittest. The fictionalist cannot say this. Leng must say that the mathematical theorem is true in the fiction, and its being true in the fiction is the reason that this biological trait is the fittest. The link between this truth in the fiction and the fitness of the trait is that there is an abstract structural relation between these fictional entities and the bees.

Just like Elgin, it remains quite unclear what sort of explanation this is. Any attempt at clarification risks offering a nonmathematical explanation or else failing to offer an explanation at all. For example, Leng could say that the fictional truth of the mathematical theorem entails a family of real–world counterfactuals: if the bees had built the honeycombs using different shapes, they would have used more wax. These counterfactuals may very well be true of the real world. It remains to be seen, though, whether it is these counterfactuals that figure in the explanation, or whether it is the link between the mathematics and the counterfactuals that explains. If it is the former, then the mathematics is merely helping us to learn the explanation. The explanation is not itself properly mathematical, and Leng is engaged in some kind of "hard road" project of removing mathematics from scientific explanations. This is not her intention. On the latter option, we again confront a mysterious link between a fictional claim and a real world claim. The connection between the fictional claim and the counterfactual is no clearer than the earlier link to the fitness of some biological trait.

I conclude that fictionalists who offer mathematical explanations while denying mathematical truth must do better if they are to respond to Colyvan's challenge. If explanation requires truth, then possessing an explanation requires indicating what is true. A fictionalist invokes a family of claims which are deemed to be false. The fictionalist thus cannot claim to have presented an explanation until they indicate where the true lies.

3.4 Naturalizing the A Priori

The proposal outlined here for the philosophical significance of mathematical explanations in science might seem dangerously unstable. A scientific realist who accepts the legitimacy of IBE is able to use these mathematical explanations to justify some purely mathematical truths. This justification is available because of the link between explanation and truth as well as the special explanatory power that mathematical truths provide. However, the way that mathematical truths explain is via their abstract structure. At the end of Section 3.2 I argued that this means that there is no IBE argument that will characterize the natures of mathematical objects. The problem is how to accept the truth of some claim about some mathematical domain without also taking a position on the intrinsic natures of these objects.

As I see it, there are only two sorts of solutions to this problem. The Quinean naturalist permits no other means of justifying claims. So they should admit that they cannot know the natures of mathematical objects and defend a kind of agnostic position about their character. If the Quinean is truly neutral on these issues, then they cannot draw a platonist conclusion.[6] The platonist position requires the existence of abstract objects which lack causal powers and reside outside of space and time. But these features are not implicated in the mathematical explanations deployed in science. All that is exploited are structural relations, and these can obtain if mathematical entities are construed in nominalistic terms. The Quinean naturalist has no way of ruling these interpretations out of bounds.

The second sort of solution is to allow some means for justifying one specific interpretation of purely mathematical claims. Quine tried to do this using his theoretical virtues, but our discussion in section I highlighted how unconvincing this is to the determined nominalist. One source of justification associated with traditional platonism is a priori justification. That is, we may have the capacity to support certain claims about mathematical entities independently of whatever course of experience we have had. If this capacity exists, then an IBE argument for the truth of some mathematical claims could be supplemented with an a priori argument concerning the natures of the subject-matter of these claims. On one view, for example, mathematics is about abstract structures. This platonist or "ante rem" structuralism is partly motivated by the role of mathematics in science, but it is also supported by the way pure mathematics has developed over time (Shapiro, 1997). It remains to be seen what claims about mathematical structures must be a priori justified in order for ante rem structuralism to be vindicated.

By way of conclusion, we can briefly consider how the admission of a priori modes of justification can cohere with a naturalistic approach to mathematics more generally. Casullo has argued for two controversial claims about a priori justification that bear on this issue (Casullo, 2003). First, a priori justification need not confer certainty. Justification in general can obtain independently of certainty, and so it is unfair to require certainty for the special case of a priori justification. Second, the existence of a priori justification is best shown through an empirical investigation of human agents. A priori arguments for the existence of a priori justification are not convincing and fail to make contact with the reliability of human reasoning. If a priori sources are to be found, then we need empirical studies of how humans reason that focus on the reliability of this form of reasoning. A promising place to begin would be with an investigation of reasoning in those areas that have been traditionally linked to the a priori such as logic, mathematics and ethics.

A naturalist should be open to the investigation and discovery of these a priori sources because the naturalist is willing to let scientific investigations shape their philosophical commitments. In this respect, both Quine and Neurath forced an overly dogmatic frame on their naturalistic approach to the philosophy of science. Neurath can be seen to err in presupposing that all traces of metaphysics would vanish from the properly streamlined ship of science. Quine mistakenly supposed that all justification is empirical and tied to the confrontation between whole scientific theories and experience. Both thus conspired to block the possibility that a source of a priori justification would be sufficient to characterize the metaphysical natures of abstract objects such as mathematical objects. The true legacy of investigations into mathematical explanations in science may thus turn out to be a revised, more liberal form of naturalism that avoids "first philosophy" while also leaving open a richer form of interaction between mathematics and science.

Notes

1 See Colyvan (2015) for a very thorough survey of these debates. Azzouni (2004) questions Quine's criterion for ontological commitment. Field (1989) and Maddy (1998) contest the claim that theoretical virtues favor a platonist interpretation of our best scientific theories.
2 Baker (2009) and Saatsi (2011) discuss many of these vexing issues. Saatsi (2016) and Baker (2016) are two important recent contributions to the debate.
3 See Mayo (1996) for another discussion of the atoms case. I have provided a preliminary comparison of the atoms case with the mathematics case in Pincock (2012, chapter 10).

4 Some platonic entities may be more easily replaced by nominalist surrogates. For example, a commitment to finitely many natural numbers may be reinterpreted in terms of concrete objects, while a commitment to infinitely many natural numbers is much harder to reinterpret. See Pincock (2012, chapter 10) for some additional discussion.

5 This objection is somewhat similar to Sober (1993). However, Sober goes further and argues that not even mathematical truth is legitimized by these considerations.

6 For additional discussion of Quine's views on ontology, see Hylton (2007, chapter 9).

References

Ayer, A. J. (Ed.). (1966). *Logical positivism*. Los Angeles: The Free Press.

Azzouni, Jody. (2004). *Deflating existential consequence: A case for nominalism*. New York: Oxford University Press.

Baker, Alan. (2009). Mathematical explanation in science. *The British Journal for the Philosophy of Science*, *60*(3), 611–633.

Baker, Alan. (2016). Mathematics and explanatory generality. *Philosophia Mathematica*, *25*(2), 194–209.

Brusatte, Stephen. (2015). What killed the dinosaurs. *Scientific American*, *313*(6), 54–59.

Brusatte, Stephen L., Butler, Richard J., Barrett, Paul M., et al. (2015). The extinction of the dinosaurs. *Biological Reviews*, *90*(2), 628–642.

Casullo, Albert. (2003). *A priori justification*. New York: Oxford University Press.

Cato & Varro. (1934). *On agriculture*. (G. P. Goold, Ed., W. D. Hooper & Harrison Boyd Ash, Trans.). Loeb Classical Library, vol. 283. Cambridge: Harvard University Press.

Colyvan, Mark. (2010). There is no easy road to nominalism. *Mind*, *119*(474), 285–306.

Colyvan, Mark. (2015). Indispensability arguments in the philosophy of mathematics. In Edward N. Zalta (Ed.), *The Stanford Encyclopedia of Philosophy (Spring 2015 Edition)*. URL =<https://plato.stanford.edu/archives/spr2015/entries/mathphil-indis/>

Elgin, Catherine Z. (2009). Exemplification, idealization, and scientific understanding. In Mauricio Suárez (Ed.), *Fictions in science: Philosophical essays on modeling and idealization* (pp. 77–90). New York, London: Routledge.

Field, Hartry (1989). *Realism, mathematics, and modality*. Oxford: Blackwell.

Hales, Thomas C. (2000). Cannonballs and honeycombs. *Notices of the AMS*, *47*(4), 440–449.

Hylton, Peter. (2007). *Quine*. New York: Routledge.

Leng, Mary. (2010). *Mathematics and reality*. Oxford: Oxford University Press.

Leng, Mary. (2012). Taking it easy: A response to Colyvan. *Mind*, *121*(484), 983–995.

Maddy, Penelope. (1998). *Naturalism in mathematics*. New York: Oxford University Press.

Mayo, Deborah G. (1996). *Error and the growth of experimental knowledge*. Chicago: University of Chicago Press.

Nahin, Paul J. (2007). *When least is best: how mathematicians discovered many clever ways to make things as small (or as large) as possible*. Princeton: Princeton University Press.

Neurath, Otto. (1934). Protocol sentences. Reprinted in Ayer (1966).

Pincock, Christopher. (2012). *Mathematics and scientific representation*. New York: Oxford University Press.

Putnam, Hilary. (1971). Philosophy of logic. Reprinted in *Mathematics, Matter and Method: Philosophical Papers Vol 1* (Cambridge: Cambridge University Press, 1979), pp. 323–358.

Quine, W. V. (1948). On what there is. *The Review of Metaphysics* 2(5): 21–38.

Quine, W. V. (1960). *Word and object*. Cambridge: MIT Press.

Quine, W. V. (1975). Five milestones of empiricism. Reprinted in *Theories and things* (Cambridge: Harvard University Press, 1981), 67–72.

Quine, W. V. (2004). *Quintessence: Basic readings from the philosophy of W. V. Quine*. (Roger Gibson, Ed.). Cambridge: Harvard University Press.

Quine, W. V., & Ullian, Joseph. (1978). *The web of belief* (2nd ed.). New York: Random House.

Saatsi, Juha. (2011). The enhanced indispensability argument: Representational versus explanatory role of mathematics in science. *The British Journal for the Philosophy of Science, 62*(1), 143–154.

Saatsi, Juha. (2016). On the 'indispensable explanatory role' of mathematics. *Mind, 125*(500), 1045–1070.

Schulte, Peter, Alegret, Laia, Arenillas, Ignacio, et al. (2010). The Chicxulub asteroid impact and mass extinction at the Cretaceous–Paleogene boundary. *Science, 327*(5970), 1214–1218.

Shapiro, Stewart. (1997). *Philosophy of mathematics: structure and ontology*. New York: Oxford University Press.

Sober, Elliott. (1993). Mathematics and indispensability. *The Philosophical Review, 102*(1), 35–57.

van Fraassen, Bas C. (1980). *The scientific image*. Oxford: Clarendon Press.

van Fraassen, Bas C. (1989). *Laws and symmetry*. Oxford: Clarendon Press.

4 Mathematical Explanation Doesn't Require Mathematical Truth

Mary Leng

According to recent explanatory versions of the Quine-Putnam indispensability argument for mathematical realism (QPIA), we have reason to believe in mathematical objects since mathematics plays an indispensable explanatory role in some of our best explanations of empirical phenomena. Some nominalists respond to this challenge by rejecting the indispensability of mathematics to these explanations. However, I agree with Christopher Pincock that in at least some cases (the honeycomb example being one), mathematics does do genuine explanatory work. I also agree with Pincock that mathematical explanations explain by picking out structural features of physical systems. Where I disagree, however, is the assumption that structural explanations that make use of mathematical theories to explain physical phenomena require those theories to be true (in the sense of consisting of bodies of truths about a domain of abstract objects). So, I will argue, mathematical explanation doesn't require mathematical truth.

As Pincock explains, the original QPIA depends on a naturalist commitment to look to our best empirical scientific theories to determine our ontological commitments. In Quine's view, our best scientific theory – that being, as Pincock puts it, "the regimented theory that maximizes the theoretical virtues" – quantifies over platonic mathematical entities, and is thus ontologically committed to such entities. So if naturalism requires us to believe our best scientific theory – at least "as a going concern" (Quine, 1975, p. 72) – then it looks as though naturalism requires us to believe in mathematical objects.

According to Pincock, there are two points of attack for nominalists in the naturalist tradition. "They may revise or deny Quine's test for the ontological commitments of a scientific theory," or "question the application of the Quinean theoretical virtues," arguing that the best theory does not quantify over abstract objects. The latter approach is that pursued by Hartry Field (2016), who wishes to show that we can formulate nominalistically acceptable versions of our ordinary 'platonistic' (or 'mathematically stated') scientific theories, and that these versions are preferable to platonistic interpretations. In particular, Field argues that his nominalistic version of Newtonian gravitational theory is preferable to the standard platonistic alternative in that it is able to provide *intrinsic explanations* of physical phenomena (that appeal only to causally relevant features), rather than the extrinsic explanations provided by their platonistic counterparts.

If in explaining the behavior of a physical system, one formulates one's explanation in terms of relations between physical things and numbers, then the explanation is what I would call an extrinsic one. It is extrinsic because the role of the numbers is simply to serve as labels for some of the features of the physical system: there is no pretence that the properties of the numbers influence the physical system whose behavior is being explained.

(Field 1985/1991, pp. 192–193)

Field's contention is that explanations that are formulated in terms of the relation between physical and mathematical objects cannot be fundamental. The mathematical objects posited in these explanations are simply serving to enable us to represent, or index, relevant features of the physical system, and it is these features that are doing all the genuine explanatory work. Nominalistically stated alternatives to ordinary platonistic scientific theories are preferable because their explanations, appealing only to physical features of physical systems, pick out the genuinely explanatorily relevant features.

Many philosophers, even on the nominalist side, are sceptical about the prospects for completing Field's nominalization project of finding nominalistically stated alternatives to our usual platonistic scientific theories, even if they would agree that, should such alternatives be found, the explanations they provide of physical phenomena would be preferable to explanations couched in mathematical terms. Pincock suggests that the alternative for nominalists who are sceptical of Field's attempts to dispense with mathematics is to revise or deny Quine's account of ontological commitment. In fact, there is some equivocation in the literature on what is meant by 'ontological commitment.' We can talk about the ontological commitments of a theory, those being the objects that would have to exist in order for the theory to be true. This is the usage at work when Quine presents his criterion of ontological commitment: "A theory is committed to those and only those entities to which the bound variables of the theory must be capable of referring in order that the affirmations made in the theory be true." (Quine, 1948, p. 33) But we also sometimes talk about *our* ontological commitments in making use of a theory – the objects we would have to commit to believing in to make our use of the theory in question reasonable (see, for example, Mark Colyvan's (2001, p. 11) presentation of the indispensability argument, whose first premise states that "*We ought to have ontological commitment to all and only those entities that are indispensable to our best scientific theories*"). With these two senses of 'ontological commitment' at work, the proposal to 'revise or deny' Quine's account of ontological commitment, as Pincock puts it, can be taken in at least two ways.

Quine's assumption is that *our* commitments in making use of a theory should line up with the theory's ontological commitments, stressing, as Hilary Putnam (1979, p. 347) puts it, "the intellectual dishonesty of denying the existence of what one daily presupposes." Some nominalists accept this alignment, but challenge Quine's account of the ontological commitments of our discourse. For example, Jody Azzouni (2004) challenges Quine's reading of the existential quantifier as ontologically committing, arguing instead that the commitments of our theories should be to objects corresponding to theoretical posits that are suitably *ontologically independent* of us. Others, though, myself included (Leng, 2010a), accept the onto-logically committing reading of the existential quantifier *literally construed*, agreeing that the literal truth of theories that existentially quantify over numbers and electrons requires the existence of such objects in their domain of quantification. In such cases, the platonistic commitments of our scientific theories are resisted by denying that *we* are committed, by our successful use of such theories, to believing those theories to be literally true (and therefore to adopting their ontological commitments as our own). As I have defended the latter approach, I will focus on this here, though I suspect that the major differences between Azzouni's views and mine lie primarily in emphasis rather than substance.

According to my fictionalist approach to mathematics and to the use of mathematics in empirical science, in formulating our empirical scientific theories we do so under the supposition that there are mathematical objects satisfying the assumptions of our mathematical theories, over and above any physical objects, and formulate our theories with that supposition in place, *without regard to whether that supposition is true*. In particular, to enable us to apply mathematics we make use of the theory of ZFU, that is, ZFC set theory with urelements, taking physical objects as urelements (nonsets that can themselves be members of sets). Once we have made the assumption that physical objects can be collected into sets, we can start applying mathematics to the physical world – making use of, for example, functions mapping sets of physical objects to associated real number quantities (their masses). For example, when we say that the mass of the Earth is 5.972×10^{24} kg, the literal content of this is to say that the earth is related by the 'mass in kilos' function to the real number 5.972×10^{24}. However, the fundamental *nominalistic* features that make this claim an appropriate one to make against the backdrop of ZFU are facts about mass-relations between the Earth and a standard kilogram unit, which are themselves unmediated by real numbers. (The key feature in this case being the fact that, if you take 5.972×10^{24} objects standing in the 'same mass as' relation to a standard kilogram, then the resulting mereological sum itself stands in the 'same mass as' relation to the Earth.) It is inconceivable that we could actually represent this relation using only the quantifiers (though this is something that is expressible 'in principle'), let alone make use of a claim so-expressed in doing any meaningful science, hence the immense value of ZFU and the mathematics of real numbers that allow us to express these underlying facts in terms of the relations to the real numbers that our mathematical theory supposes to exist. On all this, the platonist will agree, but will add that the confirmation that the resulting theory receives from its empirical successes confirms its set theoretic assumptions as true. As a fictionalist, on the other hand, I argue that the truth of the set theor-etic assumptions – interpreted literally as claims about the existence of a realm of abstract objects – is irrelevant to the success of a theory of this kind. What the set theoretic assumptions enable us to do is to express efficiently claims about physical objects and their relations that we may not otherwise be able to express or process. In particular, the fictionalist claims, what is expressed is that the physical world is configured the way *it* would have to be in order for our platonistically expressed scientific theory to be true, while remaining agnostic about the existence of the additional mathematical objects posited by that theory.

To a great extent, the motivation behind this proposed understanding of our empirical scientific theories is the same as Field's. In both cases, ordinary (platonistic) scientific theories are viewed as useful fictions, with mathematical assumptions (particularly the axioms of ZFU) providing a valuable expressive tool, enabling us to represent features of physical systems efficiently by appeal to the relations holding between appropriately related mathematical posits (such as associated real numbers). In both cases it is held that empirical scientists can and should continue to use platonistic scientific theories, without adopting the platonistic ontological commitments of those theories, but instead believing only the nominalistic content of the theories. The key difference is that, in Field's case, scientific realism (understood as belief in the truth of our *best* scientific theories) is still vindicated, as although he proposes that we adopt an instrumentalist attitude to the platonistic commitments of the platonistic scientific theories that we actually use, this attitude is justified by appeal to the existence of nominalistically stated alternatives to those theories, which we believe to be *better theories* than the ones we generally use, and whose truth would explain the instrumental success of their platonistic counterparts (as conservative extensions of these theories). The so-called 'easy road' nominalism of myself and others avoids the 'hard road' of finding nominalistically stated alternatives to our ordinary scientific theories (the labels 'easy road' and 'hard road' are due to Colyvan, 2010), but at the cost of giving up on ordinary scientific realism (understood as belief in the truth or approximate truth of our best scientific theories). Instead 'easy roaders' propose a weaker 'nominalistic scientific realism' (Balaguer, 1998, p. 130), which consists in belief in the nominalistic content of our ordinary scientific theories, but not in their platonistic components, acknowledging that the best or even only way of expressing that nominalistic content may well have to appeal to mathematical posits.

Easy road nominalists, then, believe that our best scientific theories – the ones most brimming with the Quinean theoretical virtues – are likely to be platonistic. As such, one might expect these nominalists to be immune to challenges arising out of the alleged superior explanatory power of platonistic theories as compared with nominalistic alternatives. Since they have already conceded that our most virtuous theories – including those with greatest explanatory power – are likely to be couched in platonistic terms, one might expect them to be unmoved by appeals to the explanatory virtues of platonistic scientific theories. To some extent this is right: that a scientific explanation is formulated in mathematical terms is by itself no particular reason for concern. After all, easy roaders have conceded that mathematics may well be so useful as to be *indispensable* in expressing the nominalistic content of our theories – there may simply have no better means of expressing what we take to be the facts about the physical world except by saying that the physical world is the way it would have to be in order for our mathematically stated scientific theories to be true. The indispensability of mathematical posits in formulating our best explanations of empirical phenomena should not be particularly surprising to easy-roaders, who have already conceded the vast expressive power afforded by the framework of ZFU.

This, though, hides a central difficulty. The key reason why easy roaders claim that we are not ontologically committed to the mathematical objects posited by our best scientific theories is that these posits make their way into our scientific theories in order to enable us to express what are fundamentally *nonmathematical* facts about physical objects and their relations, and this is something that can be done regardless of whether there really are any mathematical objects satisfying the assumptions of our mathematical theories. The difficulty arises when one looks at purported mathematical explanations of physical phenomena. If easy roaders wish to claim that the sole contribution made by mathematics in our scientific theories is to express some nominalistic content, then they will have to claim that the sole contribution made by mathematics *in our scientific explanations* is likewise to express some nominalistic content. As such, although mathematical assumptions may be indispensable in formulating our explanations, it looks like any genuine explanatory work that is done by these explanations must be via their nominalistic content, not their platonistic assumptions. So just as Field is committed to claiming that mathematics only plays an 'extrinsic' explanatory role, so it seems must the easy road nominalists commit to holding that, although mathematics may be indispensable to some explanations of physical phenomena, all the genuine explanatory work of such explanations resides in the nominalistic content that the mathematics is being used to express.

The difficulty here is that it looks as though there are some cases of mathematical explanations of physical phenomena where the mathematics itself is doing some genuine explanatory work, over and above any nominalistic content that the mathematics expresses. Pincock points to the Honeycomb Conjecture as an example of such a case (the example is due to Lyon and Colyvan, 2008). If we try to explain why bees build honeycombs the way they do, we can answer that this is an evolutionary adaptation that uses the minimum amount of wax. But if we push further and ask *why* this particular structure uses the minimum amount, then it looks like we have to answer that it's because its 2-D cross sections are *hexagonal*, and hexagons are the maximally efficient choice of regular polygons to use in tilings, appealing to a theorem of geometry to do the explanatory work.[1] If the only work being done by the mathematics in this explanation is to express some true nominalistic content, then it looks as though, if we were to describe the shape of the honeycomb in purely nominalistic terms, doing so should itself provide the resources for explaining why they are built this way rather than any other. But it is difficult to see how we could find any *better* understanding of what's going on this way that does not appeal to the *hexagonal* shape of the honeycomb and the efficiency of this shape over others. The mathematical explanation in this case doesn't seem to be playing proxy for some better intrinsic explanation that appeals only to fundamentally mathematics-free features of the physical situation. Rather, the mathematical features themselves seem to take central stage in providing us with an understanding of the phenomenon concerned.

I would like to agree with Pincock, then, that in this example the mathematical theorem itself, rather than any nominalistic content represented, is doing some genuine explanatory work. This is in contrast with many nominalists, who argue that in this and other cases (such as Alan Baker's well-known (2005) example of periodical magicicada cicadas, insects whose prime-number length periods of 13 and 17 years are evolutionary optimal for avoiding periodical predators) any genuine explanatory work is achieved solely by the nominalistic content that is picked out by the mathematics used (see Daly & Langford, 2009; Saatsi, 2011; and myself in a previous incarnation (Leng, 2005), for accounts along these lines). The reason why I take it that Pincock is right here is that the explanation of the phenomenon in question appears to be via the structural features of the physical system, and that such *structural explanations* are most naturally presented in mathematical terms. However, it is precisely this structural feature of such explanations that neutralizes them as a threat to nominalism. For, I will argue, mathematical explanations understood as structural explanations do not require the existence of any abstract mathematical objects or indeed structures, but only the (approximate) instantiation of axiomatically characterized mathematical theories.

What does it mean to say that mathematical explanations, such as the proposed explanation of the shape of the honeycomb by means of the Honeycomb Conjecture, are structural explanations? A structural explanation of a physical phenomenon explains that phenomenon as holding as a consequence of general (and typically mathematical) structural features instantiated in the physical circumstances to which it belongs (see, e.g., Bokulich, 2008, p. 149 for a characterization along these lines: "a structural explanation is one in which the explanandum is explained by showing how the (typically mathematical) structure of the theory itself limits what sorts of objects, properties, states, or behaviors are admissible within the framework of that theory, and then showing that the explanandum is in fact a consequence of that structure"). But how do we identify which features are to count as general structural features? In mathematics, we can do so via structure-characterizing axioms. That is, we take it that the axioms of a given mathematical theory characterize relevant structures, the (2nd order) Peano axioms for example identifying what would have to be true of any given system of objects and their relations in order for it to count as an instance of a natural number structure, and Euclid's axioms identifying what would have to be true of a given system of objects and their relations in order for it to count as an instance of a Euclidean geometry. If mathematical axioms are taken as structure-characterizing, then any consequences of those axioms can be taken to be consequences of the structure-characterizing features. To the extent that pure mathematicians are involved in drawing out the consequences of mathematical axioms, we can think of them as inquiring into structures and their features.

A natural interpretation of this picture, and one that I take it that Pincock assents to, is the platonic '*ante rem*' structuralism of Stewart Shapiro (1997) and Michael Resnik (1997). According to both, mathematics is a body of truths *about axiomatically characterized mathematical structures*, and given that our axiomatic mathematical theories include axioms that are never instantiated in physical reality, in order to preserve the truth of mathematics these structures must exist *ante rem*, that is, as abstracta over and above any actual or potential physical instantiations. But an alternative picture (such as that of Geoffrey Hellman's '*in re*' modal structuralism (1989)) takes mathematical axioms to be 'structure-characterizing' without requiring that there are any systems of objects instantiating our axioms. According to this picture, we take mathematical axioms as implicit definitions, stating what would have to be true of any system of objects in order for it to count as an instance of (say) a Euclidean geometry, without assuming that these axioms are in fact instantiated. When we derive consequences from such axioms, we can conclude that such consequences will also hold in any system of objects satisfying the axioms, *solely in virtue of the structural features* picked out by the axioms.

Hellman's modal structuralism provides an interpretation of the claims of pure mathematics as bodies of truths. In particular, in Hellman's picture, a mathematical proposition P expressed within the context of an axiomatic mathematical theory T should be interpreted not at face value, but rather as (roughly) the claim that 'The axioms of T are logically possible, and have P as a logical consequence' (which will be true in those cases where P is a theorem of a consistent axiomatic theory). Like the *ante rem* structuralist, Hellman is motivated in his structuralist account to provide an interpretation of mathematics according to which standard mathematics comes out as true. As a mathematical fictionalist, I am less concerned with preserving the truth of standard mathematics, so long as we can explain the sense in which the ordinary mathematical claims of ordinary mathematicians can be seen as *correct* if not literally true. So I reject Hellman's modal-structural reinterpretations of mathematical claims in favor of a face value reading. Nevertheless, in terms of our account of mathematics as practised, Hellman and I are on the same page: in both cases we take it that in pure mathematics (or at least, in mature, axiomatic mathematics – see Leng (2010b) for an account of pre-axiomatic theorizing) mathematicians are involved in drawing out the consequences of mathematical axioms. Thus, for my fictionalist account of pure mathematics, when a mathematical theorem is justified as being mathematically correct or 'true in the fiction,' all that is meant is that it is a consequence of the fiction–characterizing axioms.

What does this mean for the prospects for a fictionalist understanding mathematical explanations as structural explanations? Pincock complains that my account posits a 'mysterious link' between a fictional claim and a real world claim. Why, he wonders, should the fictional truth of the mathematical theorem tell us anything about the real world honeycomb? My answer to this is that the mathematical theorem is a consequence of structure-characterizing axioms (in this case, the axioms for Euclidean geometry). Those axioms are themselves *approximately* satisfied by physical points and lines on the scale that the conjecture applies. So taking a cross section of the bees' honeycomb as a plane, with nodes in the honeycomb as points and walls as straight lines, the Euclidean axioms are (approximately) true when interpreted as about this physical system. As such, Euclidean theorems are likewise (approximately) true of this physical system, including the theorem concerning the efficiency of hexagonal tilings as opposed to other regular shapes. This explanation of the bees' choice of the honeycomb shape as the most efficient use of wax thus requires no 'mysterious link' between a fictional claim and the real world, but only the straightforward interpretation of hitherto uninterpreted mathematical axioms as (approximate) truths about a physical system.

While I agree, then, with Pincock that mathematics sometimes plays an explanatory role over and above any specific nominalistic content that the mathematics is being used to represent, I do not think that *mathematical objects*, either abstract particulars (such as numbers) or abstract universals (such as *ante rem* structures) play any role in these explanations. Rather, mathematical theories provide for *structural explanations,* where a structural explanation of a given phenomenon is given whenever (a) mathematical axioms can be interpreted as (approximately) true of the relevant physical system, and (b) the occurrence of the phenomenon can be derived as a consequence of those axioms under that interpretation. Here mathematics plays an important *modal* role in convincing us that the phenomenon in question *had to occur* given the structural features of the physical system (as characterized by the mathematical axioms that are instantiated), a feature that would be hard to replicate in a mathematics-free description of the physical system that did not indicate general structural features best described by mathematical axioms. But for mathematical theories to play this role we need only for their axioms to be consistent and for their theorems to be logical consequences of those axioms, not for the axioms to be true of any structure over and above a given physical instantiation.

Note

1   While most philosophers of mathematics discussing the case have accepted that the Honeycomb Conjecture is at least relevant in explaining the shape of actual honeycombs, the example isn't universally accepted, with Tim Räz arguing that the three dimensional structure of the bee's honeycomb complicates the issue in a way that throws doubt on the relevance of the two-dimensional mathematical result. I will ignore these reservations here, in part because I take it that the conjecture is still relevant to the efficient construction of the two-dimensional cross-sections of the honeycomb, but also because, even if this explanation turns out not to be the correct one, its form as a structural explanation of structural features of the physical world is one that I believe many good explanations will take.

References

Azzouni, Jody. (2004). *Deflating existential consequence: a case for nominalism.* New York: Oxford University Press.

Baker, Alan. (2005). Are there genuine mathematical explanations of physical phenomena? *Mind, 114*(454), 223–238.

Balaguer, Mark. (1998). *Platonism and anti-platonism in mathematics.* Oxford: Oxford University Press.

Bokulich, Alisa (2008), *Reexamining the quantum-classical relation: Beyond reductionism and pluralism.* Cambridge: Cambridge University Press.

Colyvan, Mark. (2001). *The indispensability of mathematics.* New York: Oxford University Press.

Colyvan, Mark. (2010). There is no easy road to nominalism. *Mind, 119*(474), 285–306.

Daly, Chris, & Langford, Simon. (2009). Mathematical explanation and indispensability arguments. *The Philosophical Quarterly, 59*(237), 641–658.

Field, Hartry. (1991). Can we dispense with space-time? In *Realism, mathematics, and modality.* Oxford, UK; Cambridge, USA: Basil Blackwell. [Originally published in 1985, in *PSA: Proceedings of the Biennial Meeting of the Philosophy of Science Association, 2,* 33–90.]

Hellman, Geoffrey. (1989). *Mathematics without numbers: towards a modal-structural interpretation.* Oxford: Clarendon Press.

Leng, Mary. (2005). Mathematical explanation. In C. Cellucci & D. Gillies (Eds.), *Mathematical reasoning and heuristics* (pp. 167–189). London: King's College Publications.

Leng, Mary. (2010a). *Mathematics and reality*. Oxford: Oxford University Press.

Leng, Mary. (2010b). Pre-axiomatic mathematical reasoning: an algebraic approach. In Gila Hanna, Hans Niels Jahnke, & Helmut Pulte (Eds.), *Explanation and proof in mathematics*. Dordrecht: Springer.

Lyon, Aidan, & Colyvan, Mark. (2008). The explanatory power of phase spaces. *Philosophia Mathematica, 16*(2), 227–243.

Räz, Tim. (2013). On the application of the honeycomb conjecture to the bee's honeycomb. *Philosophia Mathematica, 21*(3), 351–360.

Resnik, Michael D. (1997). *Mathematics as a science of patterns*. Oxford: Clarendon Press.

Saatsi, Juha. (2011). The enhanced indispensability argument: Representational versus explanatory role of mathematics in science. *The British Journal for the Philosophy of Science, 62*(1), 143–154.

Shapiro, Stewart. (1997). *Philosophy of mathematics: structure and ontology*. Oxford: Oxford University Press.

Study Questions for Part II

1. According to Quinean naturalism, our best scientific theories require us to believe that platonic mathematical objects exist. Why? And what is Field's response to this view?
2. What is Colyvan's challenge to fictionalism? Why does Pincock think that fictionalists can't meet this challenge?
3. What is an inference to the best explanation (IBE)? Why does Pincock believe that IBE is eliminative?
4. Why does Pincock believe that mathematical explanation requires mathematical truth?
5. What is a structural explanation according to Leng? Does it require the axioms it uses to be true?
6. What is Leng's view of mathematical explanation? Does it entail that numbers exist? Why is the concept of structural explanation important for this view?

Part III

Does Quantum Mechanics Suggest Spacetime is Nonfundamental?

5   Against Wavefunction Realism

David Wallace

Wavefunction Realism: The Outline Case

The outline case for wavefunction realism begins with quantum mechanics, formulated (for N particles) something like the following:

1. The *wavefunction* of the system, at time t, is a function Ψ_t from the $3N$-fold Cartesian product of the real numbers, $\mathbf{R}^{3N}$, to the complex numbers. (We can conveniently absorb the time index into the function, so that the wavefunction assigns a complex number $\psi(x_1, \dots q_{3N}; t)$ to each $3N$-tuple of real numbers and each time t.)

2. The dynamical evolution of the wavefunction (at least when the system is not being observed) is given by the Schrödinger equation,

$$\frac{\partial}{\partial t}\psi = -\frac{i}{\hbar}\left(\sum_{k=1}^{N}\frac{\hbar^2}{2m_k}\nabla_k^2\psi + V\left(q_1, \dots, q_{3N}\right)\psi\right), \tag{5.1}$$

where m_k is the mass of the kth particle, $\hbar$ is Planck's constant, and ∇_k^2 is the Laplacian differential operator with respect to the kth triple of coordinates on $\mathbf{R}^{3N}$.

3. If the positions of the particles are simultaneously observed at time t, the probability of their $3N$ position coordinates lying in a small region of volume δV around the point $\left(q_1, \dots, q_{3N}\right)$ is

$$Pr = \left|\psi\left(q_1, \dots, q_{3N}\right)\right|^2 \delta V. \tag{5.2}$$

The third of these premises suggests that ψ ought to be thought of as some sort of probability distribution or parametrization of our restricted knowledge, but a wealth of arguments (beginning with elementary observations about interference, and proceeding through formal no-go theorems both classic (Kochen & Specker, 1967; Gleason 1957) and modern (Pusey, Barrett, & Rudolph, 2011) make clear that this strategy is not really viable, at least without concessions to operationalism

(Fuchs & Peres, 2000; Fuchs, 2002) and/or pragmatism (Healey, 2012) that philosophers of physics have by and large been loath to make. At least from a scientific realist's perspective, it looks as if the quantum state has to be taken as representational: different quantum states[1] represent different ways the world could be, not simply different levels of human information about the world.

Wavefunction realism – as advocated originally by Albert (1996) and since defended by, e.g., Ney (2013b) and North (2013) (and also by Ney, this volume) then seems to follow straightforwardly from the formalism: what the wavefunction *is*, clearly, is that familiar mathematical object, a field on space, assigning a value independently to every point in that space. But the space on which the wavefunction lives is not familiar three-dimensional space (represented mathematically by $\mathbf{R}^3$) but rather $3N$-dimensional space ('configuration space' as it is called in physics). And since N is the number of particles in the system – and since, from a metaphysician's perspective, 'the system' is presumably the whole universe, even the observable part of which contains $\sim 10^{80}$ electrons and atomic nuclei – this is a space of staggering dimension.

Quantum mechanics, then – says the wavefunction realist – reveals to us that ordinary space is illusory (Albert, 1996) or at any rate nonfundamental (Ney, 2013b). The *real* space shown to us by physics is wildly different.

The obvious questions raised by this revelation are: (i) why does space *appear* to us to be three-dimensional, and, relatedly, (ii) how do we connect this radical take on fundamental reality with our observations and so empirically confirm QM? The pragmatic response to the latter is straightforward, via the probability rule (3) above: when the wavefunction is peaked around a point in $3N$-dimensional space, we should expect to observe the particles arranged with three-dimensional coordinates corresponding to the coordinates of the point at which they are peaked. (The question of what we should observe when the wavefunction is not so peaked leads us to the related, but distinct, *quantum measurement problem*, which lies outside the scope of this chapter.) But that pragmatic response, absent additional argument, is unprincipled: *why* should we expect such observations? How does this actually follow from the fundamental ontology and dynamics?

The recent literature has largely explored two answers. The strategy of Albert, Ney, and North has sought to recover the three-dimensional world as an emergent, higher-level description, in analogy with the general pattern by which higher-level ontology emerges from the lower level. The primitive-ontology strategy of Allori et al. (2008; cf. also Allori, 2013) and Maudlin (2013) instead explicitly supplements the formalism of quantum mechanics with new, fundamental three-dimensional ontology (typically this additional ontology is also designed to solve the measurement problem).

Both strategies agree on this much: that *if* unsupplemented quantum mechanics (with or without some dynamical collapse to solve the measurement problem) is correct, then we do not fundamentally live in a three-dimensional world. But (I will argue) this is a misreading of the quantum formalism. On both technical and conceptual grounds, the move from unmodified quantum mechanics to an ontology with a single fundamental entity – the wavefunction – should be rejected.[2] Technical, because the account relies on features of a certain simplified version of quantum mechanics that does not generalize; conceptual, because even if that simplified version of quantum mechanics is accepted uncritically, the move to wavefunction realism is unmotivated.

5.2 The Technical Case against Wavefunction Realism

'Quantum mechanics,' like classical mechanics, is a framework theory: within that framework, a great many different physical systems can be described, ranging from simple two-state systems, though collections of interacting particles, through relativistic fields and even (speculatively) including the dynamics of spacetime itself. The version of quantum mechanics described above, by contrast, is a *specific* theory: to be precise, the theory of a finite number of spinless particles, interacting through long-range potential forces at nonrelativistic fields. We might call the latter theory 'toy nonrelativistic quantum mechanics,' or 'toy NRQM.' The 'toy' epithet is a little harsh from the point of view of practical physics – with tweaking, it is widely used in applications – but strictly interpreted its scope is very narrow, being insufficient to analyze (for instance) any process involving light, or any atom more complex than hydrogen. Metaphysical conclusions based on it can be of value only insofar as they generalize –either to arbitrary theories falling under the quantum framework or, at a minimum, to more realistic quantum theories with a wider domain of applicability. This is not the case, as we will see.[3]

The first thing to note is that even for toy NRQM, the particular presentation given above is only one of many mathematically equivalent ways in which the theory can be formulated. It could equally well, for instance, have been given in the *momentum space representation*, where the wavefunction obeys a quite different differential equation and where $|\psi|^2$ gives the probabilities for momentum, rather than position, measurements. Abstractly speaking, modern quantum mechanics is formulated on a (usually infinite-dimensional) space of vectors, called *Hilbert space*. Wavefunctions are just one way of describing these abstract vectors (usually called the 'position representation'); momentum-space wavefunctions are another; and there are indefinitely many more; physicists shift between them freely according to which is most helpful in solving a particular problem. And so wavefunction realism already seems to involve an arbitrary choice; though, to be fair, in toy NRQM the position representation has a central role in the formulation of the theory, so it is not difficult to imagine making a case for preferring it metaphysically.

As a first move beyond toy NRQM (one sufficient to analyze atomic structure, though not to treat phenomena involving light), consider that almost all the particles analyzed by quantum mechanics have *spin*, i.e. intrinsic angular momentum. A single particle with spin $1/2$ (the most common case, and the one that describes the electron and the proton) is described not by a wavefunction taking values in the complex numbers **C**, but by a wavefunction taking values in a two-dimensional Hilbert space – roughly, the Cartesian product $\mathbf{C}^2$ of two copies of **C**. That in itself is scarcely problematic – a complex-vector-space valued field is scarcely more metaphysically abstruse than a complex-valued field. But when we consider N particles rather than one, things get – literally – exponentially more complicated. It does not suffice to give one spin vector for each point in configuration space, or even N spin vectors. What is required, rather, is a vector in a 2^N dimensional Hilbert space, or put another way, the wavefunction for N spin-half particles is specified not by one complex number at each point of configuration space, but by 2^N complex numbers. So what wavefunction realism delivers for, say, a universe of 10^{80} spin-half particles is not 'merely' a function from a 3×10^{80}-dimensional space to the complex numbers, but a function from a 3×10^{80}-dimensional space to a $2^{10^{80}}$-dimensional complex vector space. This is a radical departure from the original conception of wavefunction realism. Or (if we are willing to choose an arbitrary direction in space, which defines a preferred set of coordinates for the $2^{10^{80}}$-dimensional space), we can instead see wavefunction realism as delivering not one complex valued wavefunction, but $2^{10^{80}}$ of them; as metaphysical underdetermination goes, indeterminacy as to whether there is fundamentally one thing or $2^{10^{80}}$ things isn't bad going.

Now let's consider dropping the 'nonrelativistic' part of toy NRQM, and considering particles with relative velocities approaching light. There is a well formulated (albeit limited) quantum theory of relativistic particles; however, that theory is formulated without direct reference to the position representation (it gives, instead, a central role to the momentum representation) and indeed there is no unproblematic way to *define* a position representation in relativistic particle physics. See (e.g.) Saunders (1998), Fleming (2000), or Halvorson (2001) for (fairly technical) details, but in essence, there are two different candidates. One has the property that it's impossible to define even in principle what it means for a particle to be sharply localized in a region; the other violates the Principle of Relativity.

From the pragmatic point of view of applying relativistic quantum mechanics, this is all harmless: the 'position' measurements we make (via cloud chambers et al.) are physical processes, and if they were ever precise enough to distinguish between different definitions of position (they're not) then their physical details would suffice to determine what is in fact being measured. But it leaves wavefunction realism without a clear definition in the relativistic regime.

This does not exhaust the problems with configuration-space representations of relativistic quantum mechanics (see Myrvold, 2015) for a thorough discussion of relativistic configuration spaces in the context of wavefunction realism). But in any case these are the least of relativistic particle mechanics' problems. In the dawn of quantum mechanics it was recognized that the theory became *inconsistent* when interactions between particles were included, unless those interactions were permitted to create or destroy particles. An interim step to address this leads to *variable-particle-number* quantum mechanics, in which the Hilbert space of the theory is the direct sum of the N-particle Hilbert space for every value of N. Wavefunction realism for a theory of this kind (even setting aside spin and the ambiguity as to what the position basis is) requires an *infinite* number of configuration spaces, and a wavefunction on each, with interactions coupling the wavefunctions on different spaces – again, the position is radically transformed.

But even variable-particle-number quantum mechanics, it turned out, failed to fully realize the lesson of relativistic interactions. That lesson led physics in due course to *quantum field theory*, in which particles themselves become emergent, high-level entities: in quantum field theory, the 'right' particle description for a given system depends upon contingent facts about that system, in particular its energy density. For instance, 'the' mass of the electron in quantum field theory is not some lawlike feature of the theory, but a parameter adjusted to best fit the details of the physical situation being modelled. More radically, although popular science often presents the proton and neutron as simple agglomerations of quarks, a more accurate gloss on quark physics is that the proton/neutron description is the most perspicuous particle description of the system at low energies and gives way to the quark description at high energies. So a fundamental ontology based on the positions of particles looks forlorn in quantum field theory.

Now, quantum field theory has an *analogy* to the position representation: in some cases ('bosonic' fields, such as the electromagnetic field), the quantum state of the field theory can be represented as a wavefunction on a configuration space — albeit a space in which the points correspond (formally) to entire instaneous configurations of a field rather than to coordinates of N particles. Such a space is infinite-dimensional (and mathematically quite badly behaved) but at least prima facie the wavefunction realist can respond to the challenge of field theory by being a realist about the field-configuration-space wavefunction. (See Ney, 2013a for an explicit case for this approach.)

My immediate feeling about this move is: if what is really intended is a wavefunction on field configuration space, shouldn't we be discussing *that* metaphysics rather than being distracted by the red herring of wavefunctions on N-particle configuration space? Granted, the latter has the virtue of being simpler to talk about, but it has the vice of being inconsistent with our current best quantum theories, which seems more serious.

But in fact even field-configuration wavefunction realism has severe technical problems. For a start, as with relativistic quantum mechanics, it is difficult to reconcile it with relativistic symmetries; indeed, making straightforward sense of it seems to require a preferred choice of reference frame. (Admittedly, most advocates of wavefunction realism are sympathetic to resolutions of the quantum measurement problem which are already in severe tension with relativity, so this may be a fairly palatable bullet for them to bite.)

More seriously, 'the' field configuration basis is often not unambiguously defined in quantum field theory. Often the same operational content may be represented through radically different choices of field: this is the phenomenon of 'duality' (for a comparatively elementary example, see Coleman, 1985, ch. 6). So the correct representation for which to express wavefunction realism is pretty radically underdetermined.

Most seriously of all, I noted that only bosonic field theories *can* be represented as wavefunctions on configuration space. Others — the 'fermionic' field theories that represent electrons and quarks (and so are central to our quantum mechanical descriptions of ordinary matter) — possess no such representation.[4]

In conclusion, wavefunction realism seems to rely on features of toy NRQM which, far from being universal features of any realistic quantum theory, drop away as soon as we generalize. At least pending very substantial technical work, we should treat with grave scepticism any suggestion that a metaphysics based on these features of toy NRQM has any real bearing on the metaphysics of our Universe.

5.3 The Conceptual Case against Wavefunction Realism

Put aside all these technical objections, and consider the metaphysics of a possible world where toy NRQM is exactly true.[5] *Even so*, the move to wavefunction realism is unmotivated.

To illustrate this, consider an example from *classical* N-particle mechanics. One natural way to formulate this theory is as follows:

- The instantaneous state of the system is represented by N points $X_1, \ldots X_N$, in three-dimensional Euclidean space.
- The dynamics is given by the differential equations

$$m_k \frac{dX_k}{dt} = \sum^{j \neq k} F_{jk}\left(X_j, X_k\right), \tag{5.3}$$

where m_k is the mass of the kth particle and F_{jk} is the force on the kth particle due to the jth.

It is extremely natural to interpret this as the theory of N particles moving in space. Notice that essentially all the structure of the world, according to the theory, is encoded in the mathematical structure of the state – specifically, in the distances between the various X_k. The actual value of a given X_k, in isolation, encodes no real information about the system: Euclidean space is featureless, with no point and no direction distinguished from another. Mathematically speaking, this is because the automorphism group of Euclidean space is the full three-dimensional group E(3) of translations and rotations: any two points are related by some translation, any two directions by some rotation.

But there is another way to represent this theory. We can define the configuration space as the product of N copies of Euclidean three-space. Each N-tuple of points $(X_1, \ldots X_N)$ now corresponds to a *single* point in this $3N$-dimensional space, and the N coupled differential equations (5.3) to a single differential equation on that space.

If we compare the theory in this formulation to the original formulation, we observe that:

1. The mathematical state is now completely featureless: a mere point. Any two states are intrinsically identical.
2. Conversely, the configuration space is much more highly structured than three-dimensional Euclidean space. Beyond mere dimension it has virtually nothing in common with $3N$-dimensional Euclidean space, and indeed is a 'space' only in the mathematician's sense, not in any sense based on a physical analogy with ordinary space.. The latter's structure is characterized by the $3N$-dimensional translation/rotation group E($3N$), whereas the symmetry

group of configuration space is simply E(3), the same as for three-dimensional Euclidean space.[6] In geometrical terms, the coordinate free structure of configuration space is most perspicuously specified via a preferred identification of each point in the former with N points in three-dimensional Euclidean space.

The move to configuration space has encoded all salient features of the system via the position of a maximally simple state (a single point) in a highly structured space, rather than via an intrinsically complex state (an Ntuple of points) located in a much less structured space. There's nothing wrong with this: it's a standard move in theoretical physics. Indeed, it's a completely general move: a state–space formalism, where all states are intrinsically identical and all structural features of the world are encoded in a state's location in a highly structured state space, can be straightforwardly defined for pretty much any dynamical theory. I draw two morals:

1. When physical theories are presented to us as formulated on spaces with much more structure than Euclidean space, we should not rush to interpret them as *physical* spaces, rather than as mathematical devices to encode information about the physical state.
2. Conversely, when a state is represented mathematically by a comparatively simple entity living in a highly structured space, we should not rush to assume that the physical world is comparably simple. Indeed, we should not rush to assume a one-to-one correspondence between mathematical states and physical entities: doing so in the case of classical mechanics would lead us to assume there is one fundamental entity, not N.

What then of quantum mechanics? The most direct analogue of the classical discussion is the *Hilbert*-space formulation of quantum theory, in which states are normalized vectors. Any two vectors are intrinsically identical (being mere lines); all the physical information about a system is encoded in the location of that vector in Hilbert space. As such, it would be naive in the extreme to be a "Hilbert-space-vector realist': to reify Hilbert space, and take it as analogous to physical space.

North (2013), in making the case for wavefunction realism, actually discusses Hilbert-space-vector realism, and of course rejects it, but her reasons are instructive: she writes that "The Hilbert space formulation seems to contain too little structure from which to construct a picture of the world as we experience it. Hilbert space does not support an objective, structural distinction between positions and other physical properties, like spin, in the way that the wavefunctions space does." But of course the Hilbert space of any particular quantum theory does indeed have enough structure to do so: any particular quantum theory is given not by a bare, unstructured Hilbert space but by that space together with an algebra of preferred observables (to which I return later). If it were not so structured, Hilbert-space quantum mechanics would be *pragmatically* unsuited to quantum mechanics, which manifestly is not the case. The reason to reject Hilbert-space-vector realism, rather, is that the space on which the Hilbert space vector is defined is much *too* highly structured to be taken as analogous to physical space: it is a state space, just as configuration space is.

The wavefunction formulation of quantum mechanics lies intermediate between the elementary formulation of classical mechanics (in which essentially all the physical structure is coded in the intrinsic properties of the state, and in which the space of the state is largely featureless) and the state-space formulations of classical and quantum mechanics (in which all the world's structure is coded via the position of the state). The wavefunction is far from featureless – a given wavefunction might be extremely highly structured – but two wavefunctions that are related by (say) an arbitrary translation or rotation on configuration space will describe radically different physics, because configuration space is also highly structured. Indeed, the structure of the wavefunction only serves to encode *quantum-mechanical* facts about a system. States corresponding to fully classical states of affairs have trivial, wave-packet wavefunctions, and again everything that distinguishes one classical state from another is encoded by the location of that wave-packet.

So reading wavefunction realism from the existence of the configuration space representation of quantum mechanics seems unmotivated, for largely the same reasons as reading Hilbert-space-vector realism from the vector representation of quantum mechanics. In both cases, the combination of the intrinsic and locational features of the state serves to encode all the physical structure of the system in question, but there is no reason to think that the formalism transparently displays anything like the appropriate metaphysical description of the system.

So what *is* "the appropriate metaphysical description"? Here's one possibility: an *N*-particle quantum state is uniquely specified by assigning a complex number to every *N*-tuple of points of space. We could perfectly well interpret these complex numbers as relational properties of *N*-tuples of spatial points, irreducible to monadic properties of individual points; that's a highly nonlocal ontology, but the phenomenology of entanglement is also highly nonlocal so that looks like a feature, not a bug. I don't want to claim that this is the clear *best* metaphysical description of quantum mechanics – I seriously doubt that it is, given the technical criticisms of the previous section – but it will do as an existence proof that there are ontologies for quantum mechanics that don't regard the highly structured configuration space as a physical space.

I have argued so far that there is no good case for simply *reading off* wavefunction realism from the quantum formalism. But can it be *argued for* as the best way to think of quantum ontology? (Here I continue to put aside purely technical objections.) North attempts to do so, but her argument is that we should prefer that ontology which has just the right level of structure to support physics, and we have seen that this fails to distinguish between wavefunction realism, Hilbert-space-vector realism and the nonlocal, three-dimensional-space based ontology of complex *N*-point relations – all are structurally isomorphic, but they are sharply different metaphysically.[7]

Ney (this volume) argues for wavefunction realism on the grounds that it delivers an ontology that is both separable and local: on wavefunction realism, the state of the whole system is given by the separate wavefunction values at each point of configuration space, and the dynamics are given by configuration-space-local equations.

The case that wavefunction realism preserves locality is somewhat unclear to me. It's true that the Schrödinger equation is configuration-space local, but if wavefunction collapse is included (as in the GRW theory) then it is nonlocal even on configuration space – and if it isn't, then we are effectively assuming the Everett interpretation, which has local dynamics even on more explicitly spatio-temporal formulations (Wallace, 2012, ch. 8). The case for separability is considerably clearer, albeit there are separable formulations of Everettian quantum mechanics on spacetime, notably that given by Deutsch and Hayden (2000; cf. discussion in Wallace & Timpson, 2010).

But in any case, we need a positive argument for why separability and locality are desirable features for our ontology. And as Ney herself persuasively argues, standard arguments for wanting these features really concern separability and locality in three-dimensional space (fundamental or emergent), and not separability or locality in a high-dimensional but phenomenologically distant fundamental space. She concludes that "[t]he case for a separable and local metaphysics for quantum mechanics then comes from more broadly philosophical considerations, special relativity, perhaps brute intuition, and additionally considerations of what provides a more coherent and stable picture."

Of these, I will pass over the "broadly philosophical considerations" and "considerations of what provides a more coherent and stable picture": ultimately they are considerations to assess when comparing wavefunction realism to other concrete ontological proposals, and those lie outside the scope of this article. The appeal to brute intuition seems problematic for reasons that Ney herself again provides: why expect our intuitive faculties, evolved as they are for the emergent classical world, to track truth about fundamental metaphysics?

A conflict between special relativity and alternatives to wavefunction realism would indeed be a strong reason in favor of the position, but I'm unsure how the conflict is supposed to go. In philosophy of spacetime, 'compatibility with relativity' usually means that the theory is formulated on Minkowski spacetime with no additional structure; in mainstream physics, it more usually means that the theory has the Lorentz group as a symmetry group. But neither definition favors wavefunction realism over alternatives. As for the former: the spacetime of wavefunction realism (configuration space × time) is not Minkowski spacetime, and in an important sense is explicitly nonrelativistic, incorporating as it does a preferred sense of simultaneity (to specify a configuration is to specify the locations of several particles *at the same time*, and the formal features required to define the relation between configuration space and Minkowski spacetime continue to include this preferred simultaneity even as the interpretation of 'configuration' space changes. As for the latter: the equations of a theory have the same (dynamical) symmetry group however they are interpreted metaphysically, and so the question of whether they are Lorentz-covariant is independent of wavefunction realism. (The answer? 'Yes' for the Everett interpretation; 'No' for most proposed relativistic generalizations of Bohmian mechanics; 'Unclear' in the case of dynamical–collapse theories.) If there is any further sense in which relativity favors wavefunction realism, it has not yet been developed.

I conclude that even if the particular features of toy NRQM that permit wavefunction realism to be formulated could be found in more sophisticated and realistic quantum theories, wavefunction realism would not be a well-motivated approach to the ontology of quantum theory: its supposed advantages are unpersuasive and the highly structured 'space' on which it is defined is not properly analogous to ordinary physical space.

5.4 Epilogue: Spacetime in Quantum Theory

So what *is* the correct ontology of quantum theory?

I don't know. The advocates of wavefunction realism certainly deserve credit for recognizing that there is a significant metaphysical question to be answered here, even if their proposed answer falls short: the ill-defined metaphysics of 'eigenvector–eigenvalue links' and 'indefinite properties' look unlikely to survive in any realist solution to the quantum measurement problem. Chris Timpson and I sketch one possibility in Wallace and Timpson (2010), and I review others in Wallace (2012, ch. 8), but these are, at best, essays in the craft; more generally, I suspect that looking for 'the' ontology of a framework theory is a category error (Wallace, 2017) and that we would do better to reformulate the question in terms of the ontology of specific quantum theories, such as the standard model of particle physics (and also to recognize that these are unlikely to be *fundamental* theories, so that hopes to learn about fundamental ontology from those theories are probably vain).

However, one striking theme in the formalism of pretty much every empirically successful *specific* quantum theory I know is that space (or rather, spacetime) concepts play a central role. In particle mechanics, the theory is normally defined in terms of the spacetime symmetry group (Galilean for nonrelativistic theories, Poincaré for relativistic): indeed, a one-particle quantum theory is frequently defined as an irreducible representation of the appropriate spacetime symmetry group (whatever the metaphysical status of that approach). In quantum field theory, the connection is significantly tighter: the theory's structure is specified via operator-valued fields, i.e. maps from spacetime points into the algebra of operators on Hilbert space. So dynamical quantities get their operational significance at least partly via their association with spacetime points. Indeed, at least according to the 'algebraic' formulation of quantum field theory, the operational significance of dynamical quantities is exhausted by their spacetime associations: a quantum field theory, in the algebraic setup, is specified completely by a map from spacetime regions to the algebra of dynamical variables associated to each region, with no further specification of which operator in a region corresponds to which physical value. Spacetime is thus mathematically required in the formulation of a quantum field theory.

Does this mean that spacetime is fundamental? It is too early to tell. Quantum field theories (which presume a fixed, background spacetime) must sooner or later give way to some quantum theory of gravity, and we do not yet have that theory, so metaphysical speculation about spacetime's status in it is probably premature. What we can say is that spacetime plays a fundamental, nonderivative role in our current best quantum theories, and so extant quantum physics gives no reason at all to expect its elimination.

Notes

1 There is a caveat: ψ and $e^{i\theta}\psi$ make precisely the same predictions and so are empirically indistinguishable even in principle. The norm in physics is to regard them as the same state (and thus to accept a slight redundancy in the formalism, which can be eliminated by moving to a density-operator or ray formulation of quantum mechanics). See Maudlin (2013) for further consideration of this point in the context of wavefunction realism.

2 Monton (2013) and Lewis (2013) also pursue a strategy somewhat along these lines, though differing significantly in the details (and focussed on the conceptual rather than the technical issues).

3 The quantum mechanics I discuss in this chapter is in all cases established work and I do not attempt to give original references.

4 A technical note: the possibility of a configuration-space representation in field theory relies on the fact that spacelike separated field operators commute, and so the collection of all such operators on a spatial hypersurface has a common set of eigenvalues. But in fermionic fields, spacelike separated operators *anticommute*. Introduction of Grassman numbers allows the *formal* introduction of something analogous to a configuration space representation, but it is at best unclear whether this has any significance beyond the purely calculational – at any rate, the burden of proof lies on the wavefunction realist here.

5 I avoid saying 'suppose that we lived in such a world': it's pretty clear that toy NRQM, in which electromagnetic radiation is wholly absent, could not support complex organisms anything like us.

6 Lewis (2013) explores this point further in his discussion of wavefunction realism.

7 Or maybe not. Those sympathetic to the ontic structural realism of Ladyman and Ross (2007), Saunders (2003) et al. – like me – might be sceptical that there is a true distinction here. But for the purposes of this chapter I assume a more straightforward metaphysics, in keeping with the presumptions of most advocates of wavefunction realism.

References

Albert, David Z. (1996). Elementary quantum metaphysics. In J. T. Cushing, A. Fine, & S. Goldstein (Eds.), *Bohmian mechanics and quantum theory: an appraisal*. Dordrecht: Kluwer.

Allori, Valia. (2013). Primitive ontology and the structure of fundamental physical theories. In Alyssa Ney & David Z. Albert (Eds.), *The wave function: essays on the metaphysics of quantum mechanics*. Oxford: Oxford University Press.

Allori, Valia, Goldstein, Sheldon, Tumulka, Roderich, et al. (2008). On the common structure of Bohmian mechanics and the Ghirardi–Rimini–Weber theory: dedicated to Giancarlo Ghirardi on the occasion of his 70th birthday. *The British Journal for the Philosophy of Science, 59*(3), 353–389.

Coleman, Sidney. (1985). *Aspects of symmetry: selected Erice lectures of Sidney Coleman*. Cambridge, UK: Cambridge University Press.

Deutsch, David, & Hayden, Patrick. (2000). Information flow in entangled quantum systems. *Proceedings of the Royal Society of London, A456*(1999), 1759–1774.

Fleming, Gordon N. (2000). Reeh-Schlieder meets Newton-Wigner. *Philosophy of Science, 67*, S495–S515.

Fuchs, Christopher A., & Peres, Asher. (2000). Quantum theory needs no 'interpretation.' *Physics Today, 53*(3), 70–71.

Fuchs, Christopher A. (2002). Quantum mechanics as quantum information (and only a little more). [ArXiv preprint at arXiv:quant-ph/0205039]

Gleason, Andrew M. (1957). Measures on the closed subspaces of a Hilbert space. *Journal of Mathematics and Mechanics, 6*, 885–893.

Halvorson, Hans. (2001). Reeh-Schlieder defeats Newton-Wigner: On alternative localization schemes in relativistic quantum field theory. *Philosophy of Science, 68*(1), 111–133.

Healey, Richard. (2012). Quantum theory: a pragmatist approach. *The British Journal for the Philosophy of Science, 63*(4), 729–771.

Kochen, Simon, & Specker, Ernst P. (1967). The problem of hidden variables in quantum mechanics. *Journal of Mathematics and Mechanics, 17*, 59–87.

Ladyman, James, & Ross, Don. (2007). *Every thing must go.* Oxford: Oxford University Press.

Lewis, Peter J. (2013). Dimension and illusion. In Alyssa Ney & David Z. Albert (Eds.), *The wave function: essays on the metaphysics of quantum mechanics* (pp. 110–125). Oxford: Oxford University Press.

Maudlin, Tim. (2013). The nature of the quantum state. In Alyssa Ney & David Z. Albert (Eds.), *The wave function: essays on the metaphysics of quantum mechanics* (pp. 126–153). Oxford: Oxford University Press.

Monton, Bradle. (2013). Against 3N-dimensional space. In Alyssa Ney & David Z. Albert (Eds.), *The wave function: essays on the metaphysics of quantum mechanics* (pp. 154–167). Oxford: Oxford University Press.

Myrvold, Wayne C. (2015). What is a wavefunction? *Synthese, 192*(10), 3247–3274

Ney, Alyssa. (2013a). Introduction. In Alyssa Ney & David Z. Albert (Eds.), *The wave function: essays on the metaphysics of quantum mechanics.* Oxford: Oxford University Press.

Ney, Alyssa. (2013b). Ontological reduction and the wave function ontology. In Alyssa Ney & David Z. Albert (Eds.), *The wave function: essays on the metaphysics of quantum mechanics* (pp. 168–183). Oxford: Oxford University Press.

Ney, Alyssa. (this volume). Separability, locality, and higher dimensions in quantum mechanics.

North, Jill. (2013). The structure of a quantum world. In Alyssa Ney & David Z. Albert (Eds.), *The wave function: essays on the metaphysics of quantum mechanics* (pp. 184–202). Oxford: Oxford University Press.

Pusey, Matthew F., Barrett, Jonathan, & Rudolph, Terry. (2011). On the reality of the quantum state. *Nature Physics, 8*(6), 476. [arXiv:1111.3328v2]

Saunders, Simon. (1998). A dissolution of the problem of locality. *PSA: Proceedings of the Biennial Meeting of the Philosophy of Science Association, 2*, 88–98.

Saunders, Simon. (2003). Physics and Leibniz's principles. In K. Brading & E. Castellani (Eds.), *Symmetries in physics: philosophical reflections* (pp. 298–308). Cambridge: Cambridge University Press.

Wallace, David. (2012). *The emergent multiverse: quantum theory according to the Everett interpretation.* Oxford: Oxford University Press.

Wallace, David. (2017). Lessons from realistic physics for the metaphysics of quantum theory. *Synthese,* 1–16.

Wallace, David, & Timpson, Christopher G. (2010). Quantum mechanics on spacetime I: Spacetime state realism. *The British Journal for the Philosophy of Science, 61*(4), 697–727.

6 Separability, Locality, and Higher Dimensions in Quantum Mechanics

Alyssa Ney

6.1 Introduction

In his paper, "On the Einstein Podolsky Rosen paradox," Bell derived a result according to which a theory capturing the statistical predictions of quantum mechanics cannot be one that avoids situations in which the result of one measurement correlates with the result of another space-like separated from it such that no prior determination could suffice to explain the correlation. He used this result to argue that:

> In a theory in which parameters are added to quantum mechanics to determine the results of individual measurements, without changing the statistical predictions, there must be a mechanism whereby the setting of one measuring device can influence the reading of another instrument, however remote. Moreover, the signal involved must propagate instantaneously, so that such a theory could not be Lorentz invariant.
>
> (1964/1987, p. 20)

As we know, it has since been observed that settings of a measuring device in one location may exhibit an instantaneous, thus superluminal dependence on outcomes in distant locations, thus confirming Bell's predictions. And so such nonlocal dependence seems to be a feature of our world, not merely for a quantum theory with "added parameters." But is this a necessary consequence of his proof and the experimental tests that we should include in our best metaphysical interpretation of quantum systems?

Although some are happy or at least resigned to accept nonlocality as a consequence of quantum theories, others seek ways to avoid it. One strategy is to reject a key assumption on which Bell's derivation of his theorem is thought to rely: the separability of quantum systems.

The goal of this chapter is to discuss what is perhaps a more promising alternative. This is to avoid both nonlocality and nonseparability by adopting a higher-dimensional interpretation of quantum systems. This higher-dimensional interpretational framework is now commonly referred to in the literature as *wave function realism* (Albert, 2013). It provides an interesting way to achieve a kind of local and separable metaphysics; however, as we will see, not all considerations in favor of locality and separability may apply to generate support for this interpretation.

6.2 Entanglement, Nonseparability, and Nonlocality

Bohm's illustration of the kind of case with which Einstein, Podolsky, and Rosen (and thus Bell) were concerned considers an extremely simple entangled state. A molecule containing two atoms is in a state in which the total spin is zero and the spin of each atom is exactly opposite to that of the other. He supposes the molecule is disintegrated by some process that does not change the total angular momentum and the atoms fly apart. One then measures the spin of the first atom. Because of the conservation of total spin, one then concludes that the spin of the second atom will be exactly opposite to that of the first (1951, p. 614).

In this EPRB scenario, our atoms are in an entangled state, the singlet state, which may be represented by the following wave function:

$$\psi_{\text{singlet}} = \frac{1}{\sqrt{2}} \mid x - up_{>A} \mid x - down_{>B} - \frac{1}{\sqrt{2}} \mid x - down_{>A} \mid x - up_{>B}.$$

The Born rule, the rule of quantum mechanics that allows us to infer probabilities for measurement results from such representations, will then tell us that were we to measure the spin states of these atoms, we would have a 50% chance of finding the first x-spin up and the second x-spin down, and a 50% chance of finding the first x-spin down and the second x-spin up.

To say that the wave function of these atoms describes an entangled state is, according to a common formulation, simply to say that they are in a state where due to some previous process or interaction, the expectation values of measurement results with respect to a particular variable are modally correlated. We are able to correctly describe a system as in an entangled state without yet getting into the metaphysics of the situation, in particular before asking whether the atoms are in a state that is either (a) nonseparable or (b) nonlocal, in the senses to be described.

6.1.1 Nonseparability

Separability is a feature of physical systems in which the systems' constituents individually occupy distinct regions of spacetime. A system located at a spacetime region R is separable when it contains subsystems located at nonoverlapping proper subregions of R and all states of the system at spacetime region R are wholly determined or grounded by the states of those subsystems. A state of such a system is a separable state when it is wholly determined by states of these subsystems.

Systems in quantum mechanically entangled states, like Bohm's pair of atoms, are often thought to exhibit fundamental nonseparability. The singlet state is thought to be an example of a nonseparable state because the atoms' being in such a spin state is not determined by any individual facts about them, including facts about their individual x-spins. For in such a state, it appears there is no definite fact about the atoms' individual x-spins. For each one, we know that if we conduct an x-spin measurement, there is a 50% chance of finding an x-up result and a 50% chance of finding an x-down result. There is nothing more definite we may say about their individual x-spins. And yet it is also a fact about the pair that if their x-spin state is to be measured, it is absolutely certain that they will be found to have opposite spins. This joint fact about the system as a whole is not determined by any fact about the atoms' individual x-spins.

Systems can be and often are in entangled states of many variables, such as spin, position, momentum, and energy. For this reason, nonseparability appears to be a pervasive feature of quantum mechanics. Below I will consider ontological interpretations of quantum systems that reveal this apparent nonseparability to be a consequence of a more fundamental metaphysics that is separable in higher dimensions. It is worth mentioning beforehand however that there is a way of ensuring separability, if one wants to adopt the Bohmian approach to quantum theories. Bohmian mechanics is an alternative approach to quantum mechanics. It contains a dual ontology of (a) particles that always possess determinate individual values of position and momentum and (b) a wave function, which is interpreted in various ways, sometimes as a physical wave in three-space, a so-called guiding wave that pushes the particles around (Bohm 1952), other times as something with more of a nomological status, determining like a law how the particles will behave over time (Goldstein & Zanghì, 2013).

In Bohmian mechanics, one may argue that there are no facts about joint states of the atoms that fail to be determined by the states of the individual atoms. There is a fact about something else, the wave function, that is not determined by the states of the atoms (taken individually or together). Bohmian mechanics interprets entanglement as a feature of states of the wave function. However, one could say the matter ontology of Bohmian mechanics is perfectly separable. Perhaps because Bohmian mechanics has this feature, it is also manifestly nonlocal.

6.1.2 Nonlocality

The issue of locality/nonlocality has frequently been conflated with that of separability/nonseparability in the scientific and philosophical literature. This is not really so surprising given the multiplicity of meanings our language assigns to 'local.' In one such usage, we may think of nonlocality as a matter of a system's features not being determined by features of its "local" (i.e. spatially more localized) parts. However, we already have a name for that feature, 'nonseparability.'[1]

Unlike separability which concerns (noncausal) metaphysical determination or grounding of the features of systems, in the sense to be discussed here, locality is a causal notion. Lange has defined locality in the following way:

> *Spatiotemporal locality*: For any event E, any finite temporal interval $\tau > 0$, and any finite distance $\delta > 0$, there is a complete set of causes of E such that for each event C in this set, there is a location at which it occurs that is separated

by a distance no greater than δ from a location at which E occurs, *and* there is a moment at which C occurs *at the former location* that is separated by an interval no greater than τ from a moment at which E occurs *at the latter location.*

(2002, p. 15)

This account captures the idea that a system is nonlocal if it manifests direct instantaneous action at a spatiotemporal distance. Thus nonlocal systems in ordinary spacetime would exhibit superluminal influence in violation of the special theory of relativity. As is standard then, we can understand locality as a principle that there is no superluminal influence, including instantaneous action across spatial distances. However, it is worth noting as it will be important below that Lange's definition is more general and does not assume what spatial background the distances δ appear in. We may ask about the preservation of locality in any spatial framework, even those less familiar than the three-dimensional spatial framework of the manifest image.

Just as entangled systems are often thought to instantiate fundamental nonseparability, so they are thought to manifest fundamental nonlocality. This was argued to be a consequence of quantum entanglement in the Einstein, Podolsky, and Rosen (EPR) paper of 1935, assuming quantum mechanics provides a complete description of reality. To make this more vivid, let's again consider the EPRB setup and imagine the atoms' spins are measured by sending them through a Stern–Gerlach apparatus, a simple device involving a pair of magnets that deflects particles in one spatial direction or another based on their spin; see Figure 6.1.

Let's imagine atom$_A$ is sent through its apparatus slightly before atom$_B$ is sent through its apparatus. The measurement on atom$_A$ will then instantaneously affect whether atom$_B$ is deflected up or down by the magnets. And this is so no matter how far apart the two Stern–Gerlach apparatuses are.

Nonlocality is widely believed to be a genuine feature of quantum systems. After the EPR paper, as we noted, Bell argued that nonlocality cannot be avoided by granting the ontological incompleteness of quantum mechanics and postulating additional variables, so even Bohmian mechanics is a nonlocal theory. Subsequently,

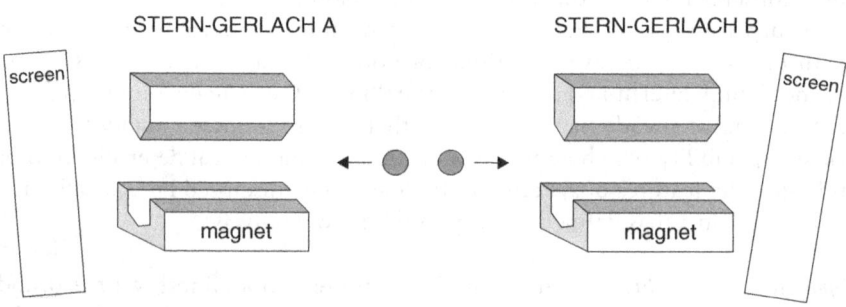

Figure 6.1

nonlocality appears to have been demonstrated in many experimental settings. The experiments by Aspect in the 1980s are perhaps the most famous (Aspect et al., 1981). However, some believe it is possible to avoid nonlocality by appealing to the nonseparability of quantum systems or by providing an even more revisionary metaphysical interpretation.

6.2 Avoiding Nonlocality

In a later section we will explore what exactly is supposed to be so problematic about nonlocality. Presently, we will address some ways of avoiding at least fundamental nonlocality in quantum mechanics. The first involves an appeal to the failure of separability. The second allows for a recovery of separability while simultaneously ensuring locality by way of a move to a higher-dimensional metaphysics.

6.2.1 Fundamental Nonseparability

Howard argues that we can ensure a local metaphysics for quantum mechanics by rejecting the separability of quantum states. He writes:

> The separability principle operates on a more basic level as, in effect, a principle of individuation for physical systems, a principle whereby we determine whether in a given situation we have only one system or two. If two systems are not separable, then there can be no interaction between them, because they are not really *two* systems at all.
>
> (1985, p. 173)

Howard's view is that when one accepts nonseparability, one thereby rejects the numerical distinctness of the individuals in entangled states. In Bohm's set-up, the proposal is we reject the belief that the atoms are distinct entities and instead view them as one. Then we can explain the observed correlations without requiring nonlocality. There is no instantaneous superluminal influence of atom$_A$ on atom$_B$ because there is no distinction between atom$_A$ and atom$_B$.

One may ask why we are entitled to conclude just by the assumption that atom$_A$ and atom$_B$ are numerically identical that there cannot be causal influence between distant wings of the experiment. Surely a state of one object can cause another state of that self-same object. My being thirsty can cause my getting up to get a glass of water. What is more plausible, however, is that causal relations are irreflexive, that one state of an individual cannot cause that very state of the self-same individual. My being thirsty at a certain time t cannot cause my being thirsty at that same time t. And so what Howard seems to be assuming is not merely that the atoms are numerically identical, but that the relevant states are identical (atom$_A$'s being found x-spin up and atom$_B$'s being found x-spin down).

Strictly speaking, this rejection of numerical distinctness of states goes beyond what nonseparability requires. We can see this by considering an alternative nonseparable approach advocated by Teller. Teller also argues that nonseparability can help us avoid nonlocality, elaborating a view he calls 'relational holism.' Teller defines this as the view that "collections of objects have physical relations which do not supervene on the non-relational physical properties of the parts" (1986, p. 73). Teller argues that cases of quantum entanglement like the EPRB setup provide genuine cases in which collections of objects have what he calls "inherent relations," relations whose instantiation are not determined by intrinsic features of the relata. They are thus genuine examples of nonseparable states.

It may not be immediately clear how Teller's view avoids nonlocality. According to relational holism, Bohm's atoms are numerically distinct, as are the states into which they enter on the two wings of the experiment. Thus it seems measurement on one atom does instantaneously affect the other some distance away. To respond, Teller (1989) grants that there will be stable correlations between the states of the two atoms in Bohm's set-up, but argues that because of the inherent relations linking them, there is no reason to infer from these correlations to a causal mechanism linking the two wings. Because of the entanglement, the correlation may be brute:

> The correlation – as an objective property of the pair of objects taken together – is simply a fact about the pair. This fact will arise from and give rise to other facts. But it need not itself be decomposable in terms of or supervenient upon some more basic, nonrelational facts. There need be no mechanism into which the correlation can be analysed.
>
> (1989, p. 222)

Thus, Teller argues relational holism allows the Bell correlations to be brute, not requiring further explanation as the consequence of a causal relation between the distant events.

It is worth noting that today, a more common nonseparable interpretation of quantum mechanics postulating irreducible relations is not framed in terms of Teller's relational holism, but rather ontic structural realism. Some versions are very much like Teller's, positing objects that bear primitive relations not explainable in terms of intrinsic features of the relata (Esfeld, 2004). Others eliminate objects altogether in favor of relations (French, 2014). I will not discuss such approaches here however since advocates of such views do not typically use structural realism in order to avoid nonlocality.

Returning to Teller, it is not clear to me why the pull to explain correlations is removed once we allow there are relations that are not determined by intrinsic features of their relata. The idea seems to be that once one gives up the assumption that *all* relational features are determined by intrinsic features of their relata (that is, in other words, once one adopts relational holism), one will thereby give up the general assumption that correlations must have explanations. Indeed he states that the adoption of relational holism frees us generally from all common cause reasoning. But even if relational holism makes it reasonable to allow *some* brute relations, it is not clear why it should remove the general presumption that correlations not be brute. To do so would seem to throw the baby out with the bathwater, giving up one of the most basic assumptions of scientific reasoning. Thus, at least for now, I would argue that prima facie, Howard's nonseparable metaphysics provides a more successful metaphysical motivation for the avoidance of nonlocality. Though, as I have noted, it involves more than a mere rejection of separability.

6.2.2 Wave Function Realism

A metaphysics for quantum mechanics that was considered and rejected early on by Schrödinger but more recently advocated by Albert (1996, 2013, 2015) has the virtue of avoiding both nonseparability and nonlocality at least at the fundamental level and perhaps simpliciter. Like Howard's interpretation, this involves the view that what appear to be distinct particles are instead manifestations of one fundamental entity. This for Albert is a single field, which he labels, following the name for the mathematical object used to represent it, the quantum wave function. It is a field in the sense that it is an object whose nature is specified by an assignment of numbers (complex values of amplitude and phase) to each point in the space it inhabits. The view is thus called 'wave function realism.' The key innovation of wave function realism is to allow that this field is not spread out in the three-dimensional space of our ordinary experience, but instead is spread out in the space in which wave functions are typically represented, a higher dimensional space. So unlike standard versions of Bohmian mechanics, in which we recognize both ordinary three-dimensional matter and a wave function, in this picture, fundamentally, there is only what inhabits the high-dimensional space of the wave function.[2]

The nature of the space the wave function inhabits is based on configuration space representations in classical mechanics. In classical mechanics, we use a configuration space of 3N dimensions to represent the possible three-dimensional locations of a system of N particles. Since classical particles always have definite locations, the locations of an entire system of N particles can be represented by a single particle at one point in configuration space. Since quantum mechanics allows individual particles to have locations that are indefinite, quantum systems will generally be represented as fields smeared out over this 3N-dimensional space. For example, in Bohm's set-up, at the start of the measurement process, the atoms will have indeterminate locations, i.e. it is indefinite whether each atom is deflected up or down by the magnetic field in its respective Stern-Gerlach apparatus. According to wave function realism, the field (the wave function) will possess nonzero amplitude at points in its space corresponding to each of these possibilities. For the nonrelativistic case, the ontology of quantum mechanics is a wave function spread over a high-dimensional space with the structure of a classical configuration space evolving according to the Schrödinger dynamics, supplemented perhaps with a collapse dynamics, depending on one's favored approach to the measurement problem.[3]

To visualize the wave function realist's proposal, we may consider first an image of how things would appear in a three-dimensional representation, when the locations of the atoms are indeterminate after they pass through their respective Stern-Gerlach devices but before wave function collapse (should there be collapse); see Figure 6.2.[4]

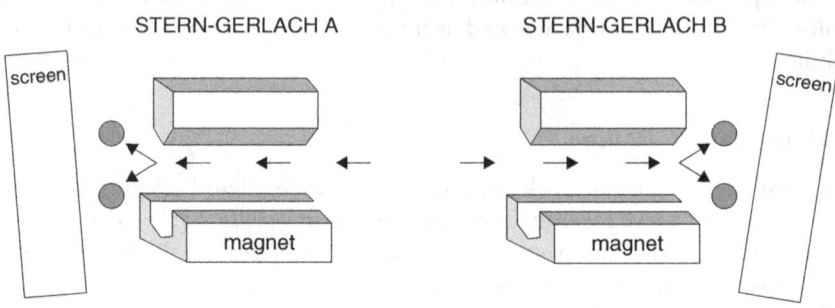

Figure 6.2

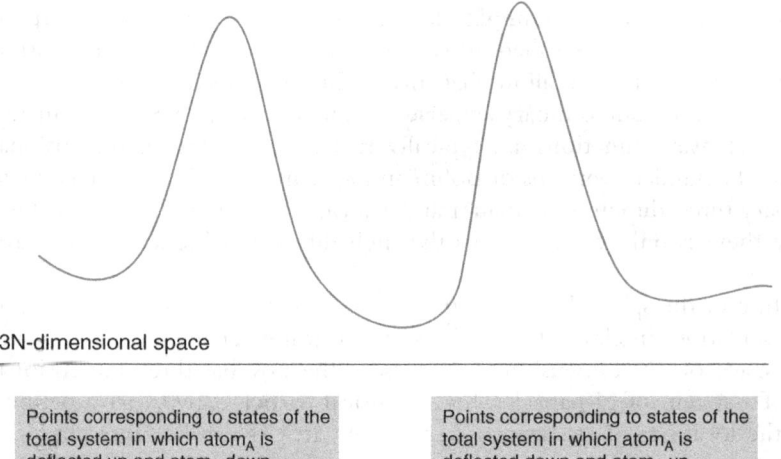

Figure 6.3

In this three-dimensional image, we do not immediately see the correlation between the first atom's being deflected upward (or downward) and the second atom's being deflected downward (or upward). We only see that each individual atom has its state spread over the two possible deflection locations. This contrasts with the higher-dimensional image; see Figure 6.3. Here each point in the higher-dimensional space corresponds to a total three-dimensional state of the entire system (states which include the positions of the atoms and the whole of the apparatus). The wave function in the case of the EPRB setup before measurement will be spread over multiple points. In particular, there are clusters of high amplitude at two regions in the higher-dimensional space, one corresponding to situations in which atom$_A$ is deflected upwards and atom$_B$ is deflected downwards, and the other corresponding to situations which atom$_A$ is deflected downwards and atom$_B$ upwards. Facts about the entanglement of the system are thus captured directly in the 3N-dimensional interpretation.[5]

The resulting wave function metaphysics is completely separable. It is separable because all states of the wave function, including the entangled states we have been considering, are completely determined by localized assignments of amplitude and phase to each point in the space of the wave function. This is not so if we want to get facts about entanglement into the three-dimensional interpretation. This requires adopting some form of nonseparability, either by denying the distinctness of atoms A and B, or as in Teller's picture, adding fundamental facts about the correlations between the atoms.[6]

Howard cites a late discussion of the EPR thought experiment in which Einstein appears to argue that if one wishes to avoid the incompleteness of quantum mechanics, one's only options are to embrace nonseparability or nonlocality (Einstein, 1949, quoted in Howard, 1985). The way Howard sees it, one can deny that objects at spatial distances have separate existences, or one must accept superluminal influence. Of course, he chooses to reject separability. What we are seeing here is that by shifting our conception of what the fundamental space is, one can avoid having to make this choice. Wave function realism provides a metaphysical interpretation in which the state of the total field is determined by its state at each point and there is no action at a distance.

To see this consequence, consider the evolution of the quantum state from a situation in which the location of the atoms (in three-space) is indefinite to one in which $atom_A$ is measured. For concreteness, assume a spontaneous collapse dynamics and the three-dimensional interpretation of quantum systems as grounded in a mass-density field as articulated by Ghirardi and Bassi (2003). What happens is that the measurement process (and resulting collapse) on the one side immediately triggers a change in the state of the system on the other side.[7] By contrast, in the higher-dimensional image, a field that is initially spread out becomes less spread out. But it is not that something on one side of the field has caused a change in the other side, causing it to become more localized. Recall that each point in the wave function's space corresponds to a total configuration of a three-dimensional system. And so it is the state of the wave function at each point at the first time that then causes the state of the wave function to be what it is at each point at the second time. What we have is simply a process in which the whole wave becomes more bunched up over time in one region of its space.[8]

The evasion of nonlocality is maintained even more clearly on a dynamics for the wave function that does not involve collapse. Again, the global state of the wave function at one time influences the global state of the wave function at all later times. What we have is simply a wave that becomes more spread out over time. Again, there is no action at a spatial distance in the fundamental metaphysics.

Now there is a question about whether there is nonseparability or nonlocality simpliciter on this picture, even if fundamentally everything is separable and local. First one must ask, can the wave function realist accept a low-dimensional metaphysics as some kind of derivative reality. This is a contentious issue we will not get into here.[9] However, if wave function realism is compatible with a derivative (but real) three-dimensional world, then there will be nonseparability in that derivative three-dimensional space, since there will be states of systems that are not determined by states of what is happening at the subregions occupied by those systems' constituents. (These states will only be determined by states of the wave function.) What about derivative nonlocality? This is a question about what explains the correlations between the spatially distant measurement events. On this view, it is not an interaction between the two wings of the experiment that explains the correlations, but instead the dynamical evolution of the wave function. So although there are certainly the observed correlations in three-dimensional space, there fails to be nonlocal influence.

And wave function realism does this in a more straightforward way than is allowed by holist approaches. I have already expressed my concern that Teller does not have a promising way of avoiding nonlocality. Although Howard may avoid what is strictly speaking nonlocal interaction between distinct objects or events, there remains on his view unexplained coordination between what appears to be happening in distant spatial regions. And so although we might not have a causal interaction between two things, but only a process involving only one, we are still committed to some mysterious coordination across distant parts of space at a time.

6.3 Why Prefer a Metaphysics for Quantum Physics that is Separable and Local?

It is possible to paint the demand for both separability and locality as the results of an unreasonable demand to make our interpretations of physical theories conform to our intuitions. For it can seem to us only natural that the properties of a whole all be traceable to, determined by properties of its parts, and also that actions do not have immediate effects across spatial distances. But, as Ladyman and Ross (2007) have rightly argued, there is no reason to believe that we would have been hardwired as a result of evolution to be good at reasoning about topics of fundamental physics or metaphysics. Our question here is what a quantum world would be like, and there is no good reason to think our intuitions are good guides to the nature of a world like this.

Some would press back on this last point. For example, Allori (2013) defends a view she finds in Einstein, that "the whole of science is nothing more than a refinement of our everyday thinking." In her view, the best physical theorizing departs as minimally as possible from the manifest image of ordinary experience, and only where it has to. But even if this is a reasonable principle, the interpretation we have seen that recovers both separability and locality also rejects a fundamental three-dimensional spatial background, replacing it with an unfamiliar, high-dimensional background. So it is not so plausible to argue that *this* separable, local metaphysics is closer to the manifest image than one that would jettison one or both of separability and locality, but retain the low-dimensional spatial background of our experience.[10]

A more promising type of argument considers the sort of interpretational assumptions that have allowed us to formulate inductively successful empirical theories. Howard considers several passages from Einstein that make this kind of case for separable and local theories. On separability, Einstein proposes:

[I]t appears to be essential for this arrangement of the things introduced in physics that, at a specific time, these things claim an existence independent of one another, insofar as these things 'lie in different parts of space.' Without such an assumption of the mutually independent existence (the 'being-thus') of spatially distant things, an assumption which originates in everyday thought, physical thought in the sense familiar to us would not be possible. Nor does one see how physical laws could be formulated and tested without such a clean separation.

(quoted in Howard 1985, pp. 187–188)

Howard speculates that what makes separability useful in the construction of successful physical theories is that it gives us a sufficient condition for the individuation of physical systems: spatial separation (Howard, 1985, p. 192). Without this, it is difficult to imagine what could provide an objective basis for individuating objects.[11]

But this doesn't seem correct. First, separability does not seem sufficient to allow for the individuation of physical systems by spatial separation. Even if we grant that all states of systems are metaphysically determined by the states of subsystems located at subregions of the region the system occupies, these subsystems (and the systems they constitute) may nonetheless fail to be clearly individuated because they possess gappy or otherwise deviant spatial trajectories. But anyway there is a more natural way to interpret Einstein's concern.

As we've seen, there are at least two ways to develop a nonseparable metaphysics for quantum mechanics. The weaker version is Teller's which would have us say that entangled systems may involve distinct entities, but that these entities instantiate inherent, i.e. irreducible relations. What this entails is that if we want to have a complete description of one of these objects that will allow us to know how it will behave over time in its environment, one must bring in facts about the other entity with which it is entangled. The facts about it and how it will behave in its environment necessarily bring in facts about the other object, which may be a significant distance away. In the absence of full knowledge of entanglement relations, this makes it challenging to predict what something in a given spatially localized set of circumstances is going to do, how it will behave. In the stronger version, Howard's, we don't know how the object will behave, for we don't even have the full object in front of us.

What can be said for locality? Again, we may bring in considerations about prediction and control. Indeed this is how Einstein appears to defend locality:

For the relative independence of spatially distant things (A and B), this idea is characteristic: an external influence on A has no *immediate* effect on B; this is known as the 'principle of local action,' which is applied consistently only in field theory. The complete suspension of this basic principle would make impossible the idea of the existence of (quasi-)closed systems and, thereby, the establishment of empirically testable laws in the sense familiar to us.

(quoted in Howard, 1985, p. 188)

If what is nearby and observable may be affected by objects that are spatially distant, then without full knowledge of the occupants of the total spacetime manifold, how are we to make predictions about how the objects we observe will behave? Locality is required to allow us to formulate testable empirical theories.

Additionally, nonlocality implies a violation of special relativity. Although the wave function realist interpretation supports nonlocal correlations in the derivative three-dimensional space of objects, it does not support nonlocality, since these correlations are not explained in terms of superluminal influence. This by itself does not mean it is compatible with special relativity – this ultimately depends on whether the theory wave function realism is an interpretation of is Lorentz covariant. Versions of quantum theory with collapse of the wave function will still include frame-dependent facts about expectation values that won't crop up in, e.g., Everettian quantum mechanics. This is an issue that runs orthogonal to the metaphysical question of wave function realism. Nonetheless, it is a virtue of wave function realism that it can avoid frame-dependent facts about causal influence.

In addition to the empirical considerations, there are also pure metaphysical considerations of coherence to consider, and some favor metaphysics that are separable. When it comes to the relational holist framework or versions of structuralism admitting relata, it is puzzling how entangled systems may be said to be constituted by distinct objects. If we are granting in the EPRB setup that one cannot give a complete specification of either one of the atoms that grounds how it will behave without bringing in a consideration of the features of the other, this suggests that the two are not fundamentally separate objects. Additionally, those sympathetic to a principle of the identity of indiscernibles (PII) have questioned whether there is a basis for individuating entangled entities. Some have used such reasoning to motivate elimination of the puzzling objects altogether (French, 2014). Howard's view avoids these difficulties by claiming that the two atoms are not distinct. But there is still a puzzle when it concerns what is inhabiting the distant regions of space. In particular, what if anything is located in the neighborhood of the individual Stern-Gerlach apparatuses? Howard and eliminative structuralists will certainly deny there is an atom on one side of the experiment or the other. But is there anything, a part of a whole or relation, on one side or the other? Admitting this would reintroduce the issue of nonlocality, but it is hard to understand how there could fail to be something in the neighborhood of an observable apparatus.

Now while the wave function realist proposal does provide a coherent, metaphysics providing clear individuation conditions and locations for its fundamental objects, it cannot reap the consequences of all of the good arguments for separability and locality considered above. Since our observations represent objects in the low-dimensional space, to have a successful physical theory, we will want to be able to make reliable predictions about what will happen when we observe or manipulate objects localized to three- or four-dimensional regions. But to the extent that wave function realism allows for the existence of a three-dimensional metaphysics, it will not be separable, and although it will strictly speaking be local, it will possess nonlocal correlations. The full separability and locality reside in the higher-dimensional picture which is unfortunately not the picture we use when we do experiments and manipulate objects.

The case for a separable and local metaphysics for quantum mechanics then comes from more broadly philosophical considerations, special relativity, and perhaps brute intuition.

6.4 Conclusion

We have seen how the higher-dimensional, wave function realist interpretation of quantum theories provides a metaphysics that is fundamentally both separable and local. If one favors separability and locality for the reasons described, then one will thereby have reason to prefer wave function realism and its attendant higher dimensions over rival interpretations of quantum theories such as standard Bohmian mechanics (which may be separable but is certainly not local) or the various holist approaches (which may be local – though I am skeptical – but are certainly not separable). It is worth emphasizing however that although the fundamental metaphysics offered by the wave function realist is both separable and local, the more pragmatic, inductive arguments favoring separability and locality are unable to provide support for this position. This is because it is only the fundamental metaphysics on this picture that is separable and local.

Notes

1 Howard (1985) is especially direct that we should avoid this conflation of concepts: "Most importantly, it should be understood that the separability of two systems is not the same thing as the absence of an interaction between them, nor is the presence of an interaction the mark of their non-separability" (p. 173).

2 The qualification "standard versions of Bohmian mechanics" is needed because Albert (1996) has also proposed a wave function realist interpretation of Bohmian mechanics, where both the matter and the wave function live in the high-dimensional space required to capture the allowable states of the wave function.

3 The total space of the wave function must actually have a more complex structure. One reason is we must include additional dimensions as well corresponding to the degrees of freedom for the spin states of the particles.

4 Note I am setting to one side Bohmian mechanics, which provides an alternative strategy for ensuring separability.

5 For much more detail on the nature of the wave function according to the wave function realist, including extension to a relativistic framework, see Ney (2013, forthcoming).

6 Loewer (1996) also argues that a move to a higher-dimensional ontology is warranted to preserve separability for quantum systems.

7 In the Ghirardi and Bassi framework, collapses are spontaneous, they are not caused by measurements. In situations describable as measurements, in which we are dealing with the interactive entanglement of a very high number of particles, the probabilities of such a spontaneous collapse event becomes extremely high.

8 Some might balk at the use of causal language here because they believe that causation has to (by definition) always involve the action of small/localized things on other small/localized things and what is happening here instead is the global state of a wave function at all points at a time determines the global state of the wave function at a

later time. But whether this is described as global causation or only determination, there is still no action at a spatial distance.

9 But see Albert (2013, 2015) and Ney (2017, forthcoming).
10 To be clear, Allori herself doesn't use this principle to argue for wave function realism.
11 Since Howard himself rejects separability, he cannot use spatial separation then as a principle for individuating physical systems. He proposes (1985, p. 198) we instead use facts about the nonexistence of quantum correlations to individuate physical systems.

References

Albert, David Z. (1996). Elementary quantum metaphysics. In J. T. Cushing, A. Fine, & S. Goldstein (Eds.), *Bohmian mechanics and quantum theory: an appraisal*. Dordrecht: Kluwer.

Albert, David Z. (2015). *After physics*. Cambridge, MA: Harvard University Press.

Albert, David Z. (2013). Wave function realism. In Alyssa Ney & David Z. Albert (Eds.), *The wave function: essays on the metaphysics of quantum mechanics*. Oxford: Oxford University Press.

Allori, Valia. (2013). Primitive ontology and the structure of fundamental physical theories. In Alyssa Ney & David Z. Albert (Eds.), *The wave function: essays on the metaphysics of quantum mechanics*. Oxford: Oxford University Press.

Aspect, Alain, Grangier, Philippe, & Roger, Gérard. (1981). Experimental tests of realistic local theories via Bell's theorem. *Physical Review Letters, 47*(7), 460–463.

Bell, John S. (1987). On the Einstein Podolsky Rosen paradox. In John S. Bell (Ed.), *Speakable and unspeakable in quantum mechanics: collected papers in quantum mechanics*. Cambridge: Cambridge University Press. [Originally printed in 1964, in *Physics Physique Fizika, 1*(3), 195]

Bohm, David. (1952). A suggested interpretation of quantum theory in terms of "hidden variables." *Physical Review, 89*(2), 166–193.

Bohm, David. (1951). *Quantum theory*. Englewood Cliffs: Prentice-Hall.

Einstein, Albert. (1949). *Albert Einstein: philosopher-scientist*. (Paul Arthur Schilpp, Ed.). Peru, Illinois: Open Court.

Einstein, Albert, Podolsky, B. & Rosen, N. (1935). Can quantum-mechanical description of physical reality be considered complete? *Physical Review* 47: 777–780.

Esfeld, Michael. (2004). Quantum entanglement and a metaphysics of relations. *Studies in the History and Philosophy of Science B, 35*(4), 601–617.

Fine, Arthur. (1996). *The shaky game*. Chicago: University of Chicago Press.

French, Steven. (2014). *The structure of the world*. Oxford: Oxford University Press.

Ghirardi, Giancarlo, & Bassi, Angelo. (2003). Dynamical reduction models. [ArXiv preprint at arXiv:quant-ph/0302164]

Goldstein, Sheldon, & Zanghì, Nino. (2013). Reality and the role of the wave function in quantum theory. In *The wave function: essays on the metaphysics of quantum mechanics*. Oxford: Oxford University Press.

Howard, Don. (1989). Holism, separability, and the metaphysical implications of the Bell experiments. In J. T. Cushing & E. McMullin (Eds.), *Philosophical consequences of quantum theory: reflections on Bell's theorem* (pp. 224–253). Notre Dame: University of Notre Dame Press.

Howard, Don. (1985). Einstein on locality and separability. *Studies in History and Philosophy of Science, 16*(3), 171–201.

Ismael, Jenann, & Schaffer, Jonathan. (2016). Quantum holism: nonseparability as common ground. *Synthese*, 1–30.

Ladyman, James, & Ross, Don. (2007). *Every thing must go.* Oxford: Oxford University Press.

Lange, Marc. (2002). *An introduction to the philosophy of physics.* Oxford: Blackwell.

Loewer, Barry. (1996). Humean supervenience. *Philosophical Topics, 24,* 101–127.

Ney, Alyssa. (2013). Introduction. In Alyssa Ney & David Z. Albert (Eds.), *The wave function: essays on the metaphysics of quantum mechanics.* Oxford: Oxford University Press.

Ney, Alyssa. (2017). Finding the world in the wave function: some strategies for solving the macro-object problem. *Synthese,* 1–23.

Ney, Alyssa. (forthcoming). The world in the wave function: a metaphysics for quantum physics. Oxford: Oxford University Press.

Teller, Paul. (1989). Relativity, relational holism, and the Bell inequalities. In J. T. Cushing & E. McMullin (Eds.), *Philosophical consequences of quantum theory: reflections on Bell's theorem* (pp. 208–223). Notre Dame: University of Notre Dame Press.

Teller, Paul. (1986). Relational holism and quantum mechanics. *British Journal for the Philosophy of Science, 37*(1), 71–81.

Study Questions for Part III

1. What is the difference between locality and separability?
2. What is wavefunction realism? How does it account for entangled systems, such as the one described in Figure 6.1 in Ney's paper, as separable and local?
3. Wallace argues that wavefunction realism is based on a simple version of quantum mechanics. Is this a problem for the view? Why?
4. What are some reasons to favor separable and local theories over nonseparable nonlocal theories?
5. Wallace argues that wavefunction realism is unmotivated. What are his responses to the motivations discussed by Ney?

Part IV

Is Evolution Fundamental When It Comes to Defining Biological Ontology?

7 Is Evolution Fundamental When It Comes to Biological Ontology? No

Maureen A. O'Malley

7.1 Introduction

There are various sorts of individuals in biology, and the two most obvious are biological and evolutionary. They share some common physical bases, but are not – in the view I advance here – the same thing. Distinguishing them is not some sort of philosophical parlour game: it is central to evolutionary and ecological analysis, as I shall show. Another way of framing the issue is to ask whether evolution has some sort of explanatory privilege over other aspects of individuality. The question motivating my chapter and Ellen Clarke's, is, therefore, phrased as: 'Is evolution fundamental when it comes to defining biological ontology?' or perhaps more aptly, ' Should we give evolution priority when defining biological ontology?' My answer is 'No,' whereas Clarke's is 'Yes.' This is not ontology for its own sake, however. As I (and Clarke) will argue, the point of worrying about ontology is because of the epistemic implications of negative or positive answers.

My discussion begins with Peter Godfrey-Smith's (2013) diagram of the relationship between evolutionary (and particularly Darwinian) individuality, and biological individuality. After outlining what each of these categories means, I go on to discuss the epistemic (explanatory and predictive) rationales for this categorization scheme. One area in which the differences between biological and evolutionary individuality can be informative is when identifying and explaining what are known as 'major transitions in evolution.' I outline how keeping distinct these two characterizations of individuals helps understand what evolutionary transitions in individuality are and how they happen.

7.2 The Distinction

Distinguishing biological from evolutionary individuals is fairly standard today, especially amongst philosophers teasing out the individuality captured by any particular biological model (e.g., Booth, 2014; Bouchard, 2011, Pradeu, 2012). Figure 7.1 represents this distinction. In the left-hand circle are things that are coherent, metabolizing, interacting entities. Because only some of these entities are thought to warrant the word 'organism,' the more general category is often labelled 'biological individuals.' For example, when considering animals such as ourselves (*Homo sapiens*, Figure 7.1: D), we usually imagine a large group of highly differentiated cells all with the same genome. We do this even though we know there are viruses and other 'foreign' mobile DNA

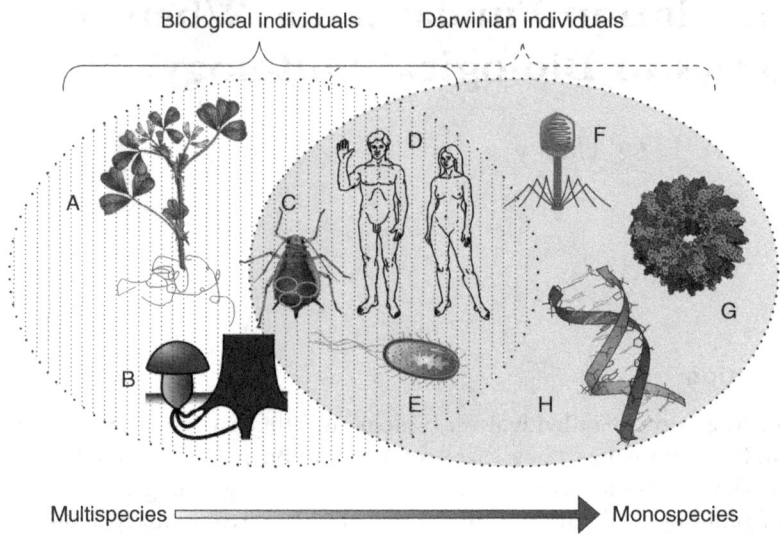

Figure 7.1 A revisualized version of Godfrey-Smith's (2013) diagram of the two kinds of individual and their relationship. A: *Medicago* and rhizobia; B: tree and mycorrhizal fungi. Mycorrhizal fungi do not have the fruiting body shown in the image. The mushroom shape is just there to indicate this is a fungal arrangement. C: aphid and *Buchnera*; D: *Homo sapiens*; E: *E. coli*; F: bacteriophage; G: tobacco mosaic virus; H: DNA (representing entities such as mobile genetic elements). Source: All images from Wikimedia Commons. Credits: A: Saint-Hilaire/ Ninjatacoshell; B: Natr; C: Snodgrass; D: Carl Sagan/NASA; E: Database Center for Life Sciences; F: Nossedotti (Anderson Brito); G: Zlir'a. H: Dcrjsr.

elements in our nuclear genome, as well as remnant bacteria in most of our cells – the mitochondria – plus all manner of functionally important microorganisms in our intestines and elsewhere. For unicellular organisms (Figure 7.1:E), we tend to consider them similarly, but as one-cell, one-genome arrangements. The biological systems that fall outside these classically 'organismal' biological individuals are of importance for how my argument develops. These are the multispecies entities (i.e., composed of organisms from different evolutionary lineages) that are illustrated in Figure 7.1 with different plant-symbiont arrangements (Figure 7.1:A and B).

In the right-hand circle are evolutionary individuals, which Godfrey-Smith (2013, 2009) has rebranded as 'Darwinian individuals.' On the far right are entities that form lineages, and on which selection operates, but which do not in most accounts have organismal status. There is notable overlap in the middle, and this is where anything that is a biological individual is also a Darwinian individual. I discuss all these distinctions in more detail in the following section; for now, note that my chapter ranges across Figure 7.1, whereas Clarke's chapter and her other work (e.g., Clarke 2013) refer only to entities that fit in the overlapping category.[1] The final dimension of this figure to bear in mind is the continuum of multispecies-monospecies running from left to right. All the boundaries in this figure are rather fuzzy.

7.3 Elaborating the Distinction

What are some of the implications of this diagram? The case of viruses helps illustrate what is at stake in distinguishing biological and Darwinian individuals. Viruses and bacteriophage (Figure 7.1: G and F respectively; the latter are the viruses that infect bacteria), plus mobile and nonmobile genetic elements[2] (e.g., plasmids and transposable elements are examples of the former; Figure 7.1: H) can form evolutionary lineages and be objects of selection. But they are not organisms, except in the most revisionary characterizations of organismality. Some of these revisions have been inspired lately by discoveries of giant viruses that have genomes larger than those of many reduced parasitic microorganisms (Raoult et al., 2004). These reconceptualizations suggest that the virus is the actively replicating entity that takes over a cell, thus making that cell a 'virus factory' (Claverie & Abergel, 2010; Forterre, 2010). The viral-factory cell in the act of cellular co-option is thus 'alive' in a sense the free-floating virus particle is not. I am not working with this redefinition in my discussion, but if it were to be used would conceivably place viruses on the other side of the diagram (the 'multispecies' end) when and only when in the process of being reproduced by cells.

Looking on the far left-hand side of the diagram, we can consider closely connected groups of organisms that are often mutualistic, persistent and regularly regenerated metabolic wholes without a common inheritance. In other words, these entities are composed of sub-entities that have separable biological and evolutionary individuality. Figure 7.1 illustrates this sort of entity with two partnerships often used as model systems: a legume (illustrated by *Medicago*) and the endosymbiotic rhizobial bacteria that enter legume roots (Figure 7.1: A), and a wider range of plants with endosymbiont fungi in and around their roots (Figure 7.1: B). Other well-known symbioses of this sort are the collectives forming corals, or hydrothermal vent tubeworms, or lichens, or dependently co-metabolizing consortia of microorganisms.[3] Liaisons such as these co-produce the main properties of the biological individual, including nutritional mode and means of generating energy, waste disposal, invader protection, and capacities for a range of interactions with other organisms and the environment. The tightest liaisons are mostly metabolic (an empirical fact, not a philosophical argument). However, as Godfrey-Smith has often pointed out, these co-produced entities – despite co-creating and sharing properties that form the collective entity – often reproduce separately. He calls the repeated co-production of such entities, generation after generation, 'reconstruction' – a process that does not generate a single fused lineage (Godfrey-Smith, 2015, p. 10122). A biological individual such as either of the examples in Figure 7.1 is thus 'a metabolic knotting of reproductive lineages that remain distinct' (Godfrey-Smith, 2013, p. 30).

This view fits with a general understanding of 'organism,' which is of entities that are self-maintaining, self-perpetuating lineage-formers that are units of selection in their own right (e.g., Queller & Strassmann, 2009; Pepper & Herron, 2008;

Santelices, 1999). These are the entities captured by the overlapping categories in the middle of Figure 7.1. As well as the expected organisms fitting there – from humans to bacteria – 'organisms' such as aphids and their endosymbionts (*Buchnera*) can be found (Figure 7.1: C; also Godfrey-Smith, 2013).[4] Why are they seen as the same unit and not placed on the left-hand side along with other multispecies entities? Because *Buchnera* are inherited vertically (parent to offspring) from the mother aphid, and not horizontally (from the environment) as happens for the entities farther to the left and outside the overlap. This is an important distinction that for many evolutionary theorists (including Clarke) demarcates the equivalent of a distinction in kind: between composite biological individuals that reproduce vertically, and composite biological individuals for which some constituents are repeatedly acquired horizontally.[5] The former are biological *and* Darwinian individuals; the latter at best biological (Godfrey-Smith, 2013).

Why does it matter if entities do not reproduce as one unit but rather as multiple units that contribute to a larger collective entity? It matters for evolutionary Darwinizing purposes because evolution is defined on the basis of reproductive success. The counting procedures that make up the machinery of population genetics are established on that basis. Although I think a simple definition of reproduction is tremendously problematic, and that there might in fact be a broader category of evolutionary individuals that are not necessarily Darwinian individuals but composed of them and aligned fitness-wise (see O'Malley, 2016), my qualms are not germane for the moment. We can simply acknowledge the fact that many collective entities, no matter how close and intimate their lives, still carry out localized reproduction (i.e., *not* across the whole system simultaneously, via some sort of bottleneck). This means that from a population-genetic perspective – the gold standard of evolutionary analysis – the fitness of the participants contributing to that collective must be considered separately (Mushegian & Ebert, 2015). As Godfrey-Smith puts it, 'counts are affected by assumptions about individuals' (2013, pp. 17–18), and population genetics for better or worse does its explanatory work with the separated individuals and not the collective ones. If population genetics does count the aggregated biological individuals, it can only be when they have reached the host-controlled reproducer stage (however, see below for how population genetics can decompose even these arrangements). We see, therefore, a decoupling of biological individuality (the lives lived) and Darwinian evolutionary individuality (turning on a reproductive fulcrum).

Any claim that biological individuals are 'less explanatory' than evolutionary Darwinian individuals (e.g., Clarke) would be focused exclusively on the overlapping section of Figure 7.1 (entities that are both biological and Darwinian). This can be a perfectly legitimate thing to do for certain epistemic purposes, as long as it is clear to everyone that the systems on the far left have been excluded (likewise those on the far right, because of not being 'alive'). The multispecies systems can indeed be broken down and analyzed as separate Darwinian individuals (e.g., the legume without its root endosymbionts; the rhizobia without their plant host). This is exactly how population genetics would treat them. But population genetics can also treat the mitochondrion in eukaryotes (e.g., animals, fungi, plants, protists) separately because it has experienced a different evolutionary history from that of the host cell's nuclear genome. Is that sort of separation what we want to do all the time? I suggest not. A case has been made along these lines against regarding 'selfish genes' as appropriate units of analysis (see Huneman, 2010). We need not think this strategy is wrong through and through, however. There are explanatory and descriptive aims that can only be achieved when the constituent entities of multispecies biological individuals are separated, and other aims that rule out such separation. These explanatory differences rest on the distinction between biological and evolutionary (i.e., Darwinian) individuals, and an acknowledgement that *neither* has out-and-out priority free of explanatory context. This is my more exact answer to the motivating question, plus a rejection of any suggestion that I am 'criticizing' the idea of Darwinian individuality.

7.4 Explanatory Implications

From what I have said so far, it should be clear that biological individuals are the units of ecological interaction and survival, achieved most basically by their metabolic capabilities. The explanatory focus for such entities is their persistence in and impact on the world. A Darwinian individual, on the other hand, is a unit of selection regardless of whether it can maintain itself metabolically. The explanatory focus for these entities is fitness measured as multiplication of the entities. To put the distinction very crudely, we explain biological individuals by viability, and Darwinian individuals by fecundity.[6] How does this work?

Take the legume, illustrated in Figure 7.1 with *Medicago* (a clover-like genus that includes alfalfa). This plant enters into a close relationship with rhizobial bacteria early in the plant's lifecycle. The bacteria become endosymbionts in the root hairs of the plant, and both roots and bacteria co-develop to achieve a nutrient exchange system. Without the bacteria, the plant has no access to nitrogen, essential for growth and the eventual maturity and reproduction of the plant. The bacteria can survive without the plant, but do less well. But although this is an intimate relationship that has evolved for 70 million years, the bacteria are reacquired horizontally for each generation of the plant (Denison and Kiers, 2011). The bacterial lineages that 'infect' the roots of the mother plant may be genetically different from but functionally identical to the lineages that inhabit the roots of the offspring plants. Because of these separable reproductive and evolutionary trajectories, the system – plant and bacteria – can be understood as the unit of *viability*: the entity that interfaces with the world, and has an impact on the reproductive capacities of the constituents. Nevertheless, the units of *fecundity* are the separate lineages, despite the fact that the plant is unlikely to reach reproductive maturity without the rhizobia. It is simply this distinction that the biological and Darwinian individuality concepts are carving up.

We can explain how the plant gets by in the world because it is part of a collective that also includes rhizobia. But the rhizobia become mere environmental conditions when the fitness chips for the plant are counted: the success of the plant is measured by how many more plants there are. Even though I think this kind of calculation has limitations (see O'Malley, 2016; also Bouchard, 2014), this is how standard evolutionary analyses are done. It is the Darwinian individual that is the focus of the counting, not necessarily the biological individual. Obviously, as Figure 7.1 shows, there is often enough overlap, but as far as population genetics goes, it is too bad for any multispecies collaborative as a unit when there is separation for counting purposes. If we want to explain more, and particularly to predict the viability of, let's say, hypothetically rhizobia-free legumes, then we return to the biological individual – the ecological entity that interacts as a unit. We may also think there is something evolutionarily interesting about how such collectives persist and change the world over evolutionary time (see below), but we cannot expect population-genetic counts to say very much about them as combined systems. If we want to predict what will happen to the Darwinian individual in the future, there are limitations to what can be said without returning to the biological individual, unless factors such as population size or selection co-efficients are known in advance. However, retrospectively, there are some very interesting explanations of Darwinian individuals that account for major turning points in the history of life on Earth.

7.5 Evolutionary Transitions in Individuality (ETIs)

ETIs occur when previously independent reproducers become a single reproducer and unit of selection (Szathmáry & Maynard Smith, 1995; Okasha, 2005; Clarke, 2014). What were once separate evolutionary individuals become single higher-level evolutionary individuals through 'deDarwinizing' processes (Godfrey-Smith 2009). Sometimes this can happen to a group of closely related entities, such as when single related cells become a multicellular entity and a new higher-level unit of selection (Michod 2005) Other times, multispecies collectives fuse their reproductive and evolutionary fates to become a single unit (e.g., cells with mitochondria). Serial achievements of ETIs are postulated to have formed the main or at least most theoretically interesting complexification trend in evolutionary history (Szathmáry & Maynard Smith, 1995; Sterelny, 1999). ETIs bring about hierarchical complexification rather than the more everyday 'horizontal' complexity (mere differentiation of structure – see McShea, 2015). However, even though this discussion refers primarily to Darwinian individuality, there are explanatory moments when biological individuality warrants consideration too.

The origin of the eukaryote cell (an event known as eukaryogenesis) is an example of where this distinction plays a role. Eukaryotic cells, which are large and filled with functionally specialized compartments (e.g., mitochondria, nucleus), arose from prokaryotic cells,[7] which are smaller cells that have less internal structure but more diverse metabolisms. The prokaryotic progenitor cell was very probably a member of the group of nonbacterial prokaryotes called Archaea (Bacteria and Archaea being the two ways to be a prokaryote). The exact first step of eukaryogenesis is unknown, but at some 'pre-eukaryotic' point an archaeal cell took in a bacterial cell and did not digest or otherwise destroy it. After quite a period of evolutionary adjustment in which the distinctive eukaryotic cell gradually emerged, the engulfed bacterium became the mitochondrion by giving up lots of its genes and functions to the host cell nucleus. It began carrying out basic metabolic and biosynthesizing processes for the host cell (although it stopped doing those activities in some lineages later on). The precise moment at which this archaeal cell carrying a bacterial cell became one unit of selection cannot be identified, but it indisputably happened: a cell with an endosymbiont – each with separate evolutionary fates for a long time – merged reproductive efforts, evolutionary trajectories, and 'achieved' another level of evolutionary (Darwinian) individuality. In fact, the fate of the mitochondrion, as I noted in a footnote above, is controlled by the host cell as the cell responds selectively to environmental pressures. We can see evidence of this in the different anaerobic and aerobic capacities of mitochondria-derived organelles in hosts adapted to anoxic, very low-oxygen and oxic regimes (Stairs et al., 2015; Müller et al., 2012).

However, the first point to note for eukaryogenesis is what I mentioned above: that even though such an ETI is universally deemed to have occurred, population genetics can still be applied to the mitochondrion on its own (ditto the chloroplast in plants, which is a similar case of an ETI – see O'Malley & Powell, 2016). In such analyses, theorists discuss the effects of mutational biases and population structure on organelle genome architecture and the efficacy of selection (e.g., Lynch et al., 2006; Birky et al., 1983). Nobody at any point, however, is suggesting that mitochondria are viable in their own right and could go out into the world beyond the cell and survive.[8] Similar analyses are carried out with *Buchnera*, the endosymbionts of aphids. They have lived inside specialized aphid cells (bacteriocytes) for 200 million years, and have lost large numbers of functional genes (Moran & Mira, 2001). Despite *Buchnera*'s deep dependence on the aphids, population genetics is often carried out on them to explain the forces behind their genomic and functional reduction (e.g., Moran, 1996; Woolfit & Bromham, 2003). But again, there is not the minutest possibility that these endosymbionts could be removed from the bacteriocytes and 'set free' to interface directly with the environment outside aphids. This is what their close but unreduced relatives, *Escherichia coli*, are able to do. In other words, *Buchnera* (and mitochondria and chloroplasts) are not expected to be biological individuals although they are expected to conform to the characteristics of Darwinian individuals, when being explained population-genetically. Although *Buchnera* are still regarded as quasi-organismal and not organelles (which is what mitochondria and chloroplasts are), they have a similar status in where they fall with respect to Figure 7.1.

But there is also a more radical way in which separating biological and evolutionary individuality affects the centrality of ETIs in explaining major transitions. When Darwinian individuals are the focus of such macroevolutory explanations, they invoke primarily a genetic account of those individuals (just as microevolutionary ones do). Why? Because the main way in which to decide what has continuity and cohesion reproductively (and is thus reproducing and forming lineages of the right sort) is by tracking genes. This way, different lineages bearing the appropriate genetic material are the individuals of interest (this is exactly why mitochondria, chloroplasts and *Buchnera*, which have their own genomes, can be analyzed population-genetically, and also the reason why Godfrey-Smith's exemplar in this category is 'genes' or 'chromosomes'). A great deal of ETI work has been focused on genes and genomes as the informational bearers of individuality, and on ETIs as the resolution of genetic conflict (e.g., Szathmáry & Maynard

Smith, 1995). However, we can also think about evolutionary transitions more metabolically, and ultimately explain these turning points by properties associated with biological individuality.

Recently Russell Powell and I have pushed this line, arguing that major transitions in evolution can also be identified and explained metabolically (O'Malley & Powell, 2016). We noted several overlooked candidates for major transitions: the divergence of Bacteria and Archaea (probably the most fundamental divergence in the history of life on Earth),[9] and the Great Oxidation Event in which the metabolism of cyanobacteria irrevocably changed planetary ecology and thus the evolutionary trajectory of all lifeforms and their future possibilities. While both these turning points are of the utmost evolutionary importance for understanding why life is the way it is now, these events cannot be accounted for in a major transitions framework focused on ETIs. A metabolic focus, however, immediately homes in on them (e.g., Sojo et al., 2014; Falkowski & Godfrey, 2008). In such instances, features of biological individuality play an explanatory role that is not just proximate (how) but also ultimate (why). Making that argument requires some elaboration on the points I made with Powell.

Our paper suggests that some major evolutionary transitions are very likely about genetic (Darwinian) individuals *and* metabolic (biological) individuals (e.g., the origin of cells, the origins of multicellularity). Some transitions, however, might be better explained by Darwinian individuality alone (e.g., eusociality). And quite a few of either a short traditional list (i.e., Szathmáry & Maynard Smith, 1995) or a longer revised list might really best be seen as major transitions explained ultimately by metabolism and biological individuality, rather than Darwinian individuality as tracked by genes. To understand why certain evolutionary transitions in individuality occurred, such as the acquisition of the mitochondrion, or the many acquisitions of plastids (photosynthesizing organelles such as chloroplasts, that were once independent organisms), Powell and I suggest that metabolism is not mere background to the ultimate (genetic) explanations, but is a driver of the events in which the resolution of genetic conflict (the prime explanandum in ETI accounts) is in fact secondary.

Eukaryogenesis may well be one of those transitions because metabolic explanations of this event are able to account for the details of why and how this 'fusion' progressed – in a way that tracking genetic individuals cannot (Lane & Martin, 2010; O'Malley & Powell, 2016).[10] Metabolic regulation is a plausible explanation of how conflict was resolved between the two cells that fused into the novel eukaryote; metabolic regulation may also have ongoing explanatory implications for the evolution of various forms of eukaryote multicellularity (Blackstone, 2013). The oxygenation of the Earth – an event brought about initially by the combination of two previously disconnected photosynthetic pathways in one organism – is another transition explained metabolically (Hohmann-Marriott & Blankenship, 2011; O'Malley & Powell, 2016). This world-changing biological event is necessarily left out of ETI-oriented major transition models because of not being explicable in terms of new levels of evolutionary individuality. Metabolism duly identifies and explains this major ecological and evolutionary transformation. Thus, even at a macroevolutionary level of explanation, metabolic (biological) individuals continue to be worth distinguishing epistemically from standard genetic Darwinian individuals. By all means, keep both involved explanatorily, but they should not be conflated.

7.6 The Bottom Line

There are good reasons to keep biological and evolutionary (Darwinian) individuality conceptually distinct, in particular to understand their different implications for ecological and evolutionary explanations. In the process, we may gain some interesting insights into biological individuals and their roles in ultimate and not just proximate explanation.

Notes

1 Clarke (this volume) thinks that viruses and mobile DNA elements need not be considered, because in most accounts they are not alive. Those same accounts, however, would agree that viruses are evolving and subject to selection.
2 'Genes' are examples of the latter in Godfrey-Smith's original diagram, and I will only mention in passing their problems as an exemplar for this discussion. Mitochondria and chloroplasts (plus other plastids) might be considered additional examples, but because their evolutionary fates appear controlled by the evolutionary fates of their hosts, they would not normally be given any stand-alone status in such a diagram. For further discussion of such entities, see the main text.
3 For some less well-known but just as fascinating systems of this sort, see Herron et al. (2013).
4 Haber (2013) would also place eusocial insect colonies ('superorganisms') here, even if he does not want to call them organisms.
5 I will ignore for the purposes of this chapter just how very mixed (vertical and horizontal) most transmission systems are (Ebert, 2013; see O'Malley, 2016 for further discussion).
6 Michod (2005), who also uses the terminology of 'viability' and 'fecundity,' sees these units as mutually overlapping, thus also answering 'yes' to the motivating question for this chapter.
7 There are some radical eukaryote-first speculations we can safely ignore here.
8 As a curious aside, this is what an evolution-of-endosymbiosis theorist, Ivan Wallin (1883–1969) argued he'd done in the 1920s. He believed mitochondria were autonomous bacteria – i.e., both Darwinian and biological individuals – and that he had cultured them outside the cell (Wallin, 1927). Sadly for his career he was not much believed, although his more general ideas linger on in discussions of endosymbiosis today.
9 This is especially the case if, as now seems likely, eukaryotes are just a specialized sort of Archaea.
10 This point obtains regardless of exactly when the mitochondrion was acquired – a topic that is a matter of considerable contestation at the moment.

References

Birky, C. William, Maruyama, Takeo, & Fuerst, Paul. (1983). An approach to population and evolutionary genetic theory for genes in mitochondria and chloroplasts, and some results. *Genetics*, *103*(3), 513–527.

Blackstone, Neil W. (2013). Why did eukaryotes evolve only once? Genetic and energetic aspects of conflict and conflict mediation. *Philosophical Transactions of the Royal Society London B: Biological Sciences*, *368*(1622), 20120266.

Booth, Austin. (2014). Populations and individuals in heterokaryotic fungi: a multilevel perspective. *Philosophy of Science*, *81*(4), 612–632.

Bouchard, Frédéric. (2011). Darwinism without populations: a more inclusive understanding of the "Survival of the Fittest." *Studies in History and Philosophy of Biological and Biomedical Sciences*, *42*(1), 106–114.

Bouchard, Frédéric. (2014). Ecosystem evolution is about variation and persistence, not populations and reproduction. *Biological Theory*, *9*(4), 382–391.

Clarke, Ellen. (2013). The multiple realizability of biological individuals. *The Journal of Philosophy*, *110*(8), 413–435.

Clarke, Ellen. (2014). Origins of evolutionary transitions. *Journal of Biosciences*, *39*(2), 303–317.

Claverie, Jean-Michel, & Abergel, Chantal. (2010). Mimivirus: the emerging paradox of quasi-autonomous viruses. *Trends in Genetics*, *26*(10), 431–437.

Denison, R. Ford, & Kiers, E. Toby. (2011). Life histories of symbiotic rhizobia and mycorrhizal fungi. *Current Biology*, *21*(18), R775–R785.

Ebert, Dieter. (2013). The epidemiology and evolution of symbionts with mixed-mode transmission. *Annual Review of Ecology, Evolution, and Systematics*, *44*, 623–643.

Falkowski, Paul G., & Godfrey, Linda V. (2008). Electrons, life and the evolution of Earth's oxygen cycle. *Philosophical Transactions of the Royal Society B: Biological Sciences*, *363*(1504), 2705–2716.

Forterre, Patrick. (2010). Defining life: the virus viewpoint. *Origins of Life and Evolution of Biospheres*, *40*(2), 151–160.

Godfrey-Smith, Peter. (2009). *Darwinian populations and natural selection*. Oxford: Oxford University Press.

Godfrey-Smith, Peter. (2013). Darwinian individuals. In Frédéric Bouchard & Philippe Huneman (Eds.), *From groups to individuals: evolution and emerging individuality* (pp. 17–36). Cambridge: MIT Press.

Godfrey-Smith, Peter. (2015). Reproduction, symbiosis, and the eukaryotic cell. *Proceedings of the National Academy of Sciences*, *112*, 10120–10125.

Haber, Matt. (2013). Colonies are individuals: revisiting the superorganism revival. In F. Bouchard & P. Huneman (Eds.), *From groups to individuals: evolution and emerging individuality* (pp. 195–217). Cambridge, MA: MIT Press.

Herron, Matthew D., Rashidi, Armin, Shelton, Deborah E., et al. (2013). Cellular differentiation and individuality in the 'minor' multicellular taxa. *Biological Reviews*, *88*(4), 844–861.

Hohmann-Marriott, F., Martin, & Blankenship, Robert E. (2011). Evolution of photosynthesis. *Annual Review of Plant Biology*, *62*, 515–548.

Huneman, Philippe. (2010). Assessing the prospects for a return of organisms in evolutionary biology. *History and Philosophy of the Life Sciences*, *32*, 341–371.

Lane, Nick, & Martin, William. (2010). The energetics of genome complexity. *Nature*, *467*(7318), 929–934.

Lynch, Michael, Koskella, Britt, & Schaack, Sarah. (2006). Mutation pressure and the evolution of organelle genomic architecture. *Science*, *311*(5768), 1727–1730.

McShea, Daniel W. (2015). Three trends in the history of life: an evolutionary syndrome. *Evolutionary Biology*, *43*(4), 531–542.

Michod, Richard E. (2005). On the transfer of fitness from the cell to the multicellular organism. *Biology & Philosophy*, *20*(5), 967–987.

Moran, Nancy A. (1996). Accelerated evolution and Muller's rachet in endosymbiotic bacteria. *Proceedings of the National Academy of Sciences, 93*(7), 2873–2878.

Moran, Nancy A., & Mira, Alex. (2001). The process of genome shrinkage in the obligate symbiont *Buchnera aphidicola*. *Genome Biology, 2*(12). [research0054.1-0054.12]

Müller, Miklós, Mentel, Marek, Hellemond, Jaap J. van, et al. (2012). Biochemistry and evolution of anaerobic energy metabolism in eukaryotes. *Microbiology and Molecular Biology Reviews, 76*(2), 444–495.

Mushegian, Alexandra A., & Ebert, Dieter. (2015). Rethinking "mutualism" in diverse host-symbiont communities. *Bioessays, 38*(1), 100–108.

O'Malley, Maureen A. (2016). Reproduction expanded: multigenerational and multilineal units of evolution. *Philosophy of Science, 83*(5), 835–847.

O'Malley, Maureen A., & Powell, Russell. (2016). Major problems in evolutionary transitions: how a metabolic perspective can enrich our understanding of macroevolution. *Biology & Philosophy, 31*(2), 159–189.

Okasha, Samir. (2005). Multilevel selection and the major transitions in evolution. *Philosophy of Science, 72*(5), 1013–1025.

Pepper, John W., & Herron, Matthew D. (2008). Does biology need an organism concept? *Biological Reviews, 83*(4), 621–627.

Pradeu, Thomas. (2012). *The limits of the self: immunology and biological identity*. Oxford: Oxford University Press.

Queller, David C., & Strassmann, Joan E. (2009). Beyond society: the evolution of organismality. *Philosophical Transactions of the Royal Society of London B: Biological Sciences, 364*(1533), 3143–3155.

Raoult, Didier, Audic, Stéphane, Robert, Catherine, et al. (2004). The 1.2-megabase genome sequence of Mimivirus. *Science, 306*(5700), 1344–1350.

Santelices, Bernabé. (1999). How many kinds of individual are there? *Trends in Ecology & Evolution, 14*(4), 152–155.

Sojo, Victor, Pomiankowski, Andrew, & Lane, Nick. (2014). A bioenergetic basis for membrane divergence in archaea and bacteria. *PLoS Biology, 12*(8), e1001926.

Stairs, Courtney W., Leger, Michelle M., & Roger., Andrew J. (2015). Diversity and origins of anaerobic metabolism in mitochondria and related organelles. *Philosophical Transactions of the Royal Society B: Biological Sciences, 370*(1678), 20140326.

Sterelny, Kim. (1999). Bacteria at the high table. *Biology & Philosophy, 14*(3), 459–470.

Szathmáry, Eörs, & Smith, John Maynard. (1995). The major evolutionary transitions. *Nature, 374*(6519), 227–232.

Wallin, Ivan Emmanuel. (1927). *Symbionticism and the origin of species*. Baltimore: Williams & Wilkins.

Woolfit, Megan, & Bromham, Lindell. (2003). Increased rates of sequence evolution in endosymbiotic bacteria and fungi with small effective population sizes. *Molecular Biology and Evolution, 20*(9), 1545–1555.

8 Is Evolution Fundamental When It Comes to Defining Biological Ontology? Yes

Ellen Clarke

To make a case for an affirmative answer to the title question I should say something about what ontology, especially biological ontology, is; Something about what sort of criteria we might use in evaluating different approaches to defining biological ontology; And, finally, what it would mean to take evolution as fundamental in the latter activity. Most of the discussion will center on a particular item of biological ontology – the *individual* – and on the merits of an evolutionary definition of that item, as opposed to a metabolic definition.

8.1 Ontology: What Are Concepts For?

Ontology is the study of what there is; of what sorts of things exist. It describes the attempt to come up with a classification scheme that lists the underlying furniture of reality. Biological science involves all sorts of specialist words with technical meanings. These words are supposed to help us, when we look at biological things (we can call separate things 'particulars'), to divide those things into different groupings, called 'kinds.' We place things together in the same kind when they share some properties in common. For example, 'HoxA' refers to a group of genes that are found on chromosome 7 of the human genome, as well as in many other lineages, and which are important in determining the body plan of the developing embryo. We name kinds, in this way, to help us in making inferences about things – in predicting how things are going to behave.

Living stuff can be parcelled up into many different kinds of particulars. Some possible kinds – such as those picked out by terms like 'protein,' 'cell,' 'liver,' and 'gene' – seem more obvious to us than others. But kinds are easy to come up with. Jorge Luis Borges' mock encyclopaedia divided animals into fourteen different kinds, including 'Those that belong to the emperor, embalmed ones, those that are trained, suckling pigs, fabulous ones, stray dogs' and 'Those drawn with a very fine camel hair brush' (Borges, 1937). There is a possible kind that contains the topmost half of every human's body. The kind is not empty – people really do have top-most halves. What is doubtful is how useful it could be.

The truth is, we don't want to detail all of the different kind concepts that are possible. But which ones do we want? Sometimes philosophers distinguish silly examples like mine and Borges' from 'natural kinds,' where the latter pick out groupings that are discovered in nature, rather than made up by us. The line between the two is difficult to draw, however. It is easier to agree that some kinds are more useful than others. Although usefulness is always relative to a purpose, some kinds are useful across a wider range of different purposes than others. For example, a mushroom hunter might refer to a classification of fungus in order to find out if it is edible or not (Dupré, 1993). But chemical element classifications – probably the most useful kinds we have ever named – are useful for chefs, and also paint mixers, and fireworks manufacturers and many other groups of people who have divergent purposes in classifying the properties of chemicals. In evaluating definitions of kind concepts, then, I suggest we rank the more useful concepts as more valuable – more worth holding on to, worth teaching – than alternatives that have fewer uses, or are useful across a smaller range of circumstances. This claim applies to kinds in general, but here I focus on applying it to one kind in particular – the biological individual.

I'm going to argue that evolutionary concepts of biological kinds are *more* useful than other concepts, at least in the special case of the kind 'biological individual' and of my 'Levels of selection' account of it. That is, in the particular case of the kind 'biological individual, I'll argue that the Levels of selection definition is the most useful. 'Fundamental' is a daft word really, an indirect way of putting things in caps and little more. It neighbors with 'most important,' 'most interesting,' perhaps suggests that other things can be reduced to it. I make none of those claims for the evolutionary definition of the biological individual. Yet I will certainly defend the importance and interest of the evolutionary definition. And I will even present some reasons to think that the evolutionary definition has a certain sort of priority over other definitions.

8.2 Biological Ontology: Individuals

O'Malley and I share an interest in arbitrating the usefulness of concepts aimed at picking out biological chunks which are smaller than clades but bigger than organs. Naming the chunk of interest is made complicated by the huge number of distinct concepts (see Table 8.1) which have been targeted at these units, none of which quite coincides with the concepts O'Malley and I each endorse.

In the table I've used a numbering system to indicate where there is repetition or where different definitions constitute rivals for a single concept. The two concepts with the greatest number of alternative definitions are 1 and 2, which we may think of roughly as 'evolutionary' and 'organizational' concepts respectively. This rough method indicates that there are around nine distinct concepts named in this table. The table exhausts neither possible nor actual concepts in the vicinity. Some concepts – number 6, for example, are obviously distinct and not intended to compete against the others – we might say that its resemblance to the others is only semantic. Yet all of the concepts are united in picking out some thing that is a *biological* thing, and which is supposed to conform to some very general standards of particularity (thingness), such as spatiotemporal location and cohesion.[1]

While O'Malley advocates a particular sort of organizational concept – a metabolic concept –of the individual, I advocate instead for a sort of evolutionary concept. These are distinct concepts, rather than rival definitions of a single concept.

Table 8.1

Term	Used by	To mean	Examples	
Evolutionary individual	Michod, 2005; Ereshefsky & Pedroso, 2015; Bouchard & Huneman, 2013	Unit which exhibits heritable variance in fitness	*Volvox carteri* Bacterial biofilms	1a
Evolutionary individual	Janzen, 1977	Genetic individual – all the parts share one unique genome	Dandelion clone, aphid clone	1b
Organism	Kant, 1790	Unit which exhibits organization	Horse	2a
Organism	Pradeu, 2010	Physiological individual, delimited by immune system	Human-gut microbes collection, *Botryllus schlosseri*	2b
Organism	Wilson & Sober, 1989	Unit which exhibits functional integration	Eusocial insect colonies, Squid–vibrio collection	2c
Organism/ biological individual	Godfrey-Smith, 2013			
Organism	Pepper & Herron, 2008	Evolutionary individual	Mouse, Honeybee colonies, *Buchnera*-aphid collection	1a
	Queller & Strassmann, 2009			1c
	Folse & Roughgarden, 2010			1d
Superorganism	Gardner & Grafen, 2009	Evolutionary individual	Clonal groups Honeybee colonies	1e
Darwinian individual	Gould & Lloyd, 1999	Units at *all* levels of compositional hierarchy	Gene, mitochondrion, cell, horse, species	3
Simple reproducer, Darwinian individual	Godfrey-Smith, 2009	Evolutionary individual	Bacterium	1a
Scaffolded reproducer, Darwinian individual	Godfrey-Smith, 2009	Lineage-forming part of an evolutionary individual	Virus, chromosome	1f

Table 8.1 (*Cont.*)

Term	Used by	To mean	Examples	
Collective reproducer (higher-level Darwinian individual)	Godfrey-Smith, 2009	Unit which exhibits bottleneck, germ separation and integration	Human, Aphid-*Buchnera* collection, colony, buffalo herd	1g
Biological individual	J Wilson, 1999	Biological particular	Developmental module, organ, protein, gene regulatory network	4
Unit of selection	Lewontin, 1970	Unit which exhibits heritable variance in fitness	Deer, cellular organelles,	1a
Unit of selection	Maynard Smith, 1987	Unit which exhibits fitness variance	Somatic cells	5
Unit of selection	Brandon, 1999	Developmental module	Neural crest	6
Unit of selection	Lloyd, 2005	Interactor	Horse	7
		Replicator	Gene	8
		Manifester of adaptation	Horse	1e
		Beneficiary of adaptation	Gene	9
Unit of evolutionary transition, reproducer	Griesemer, 2000	Unit which copies with material overlap and development	Giraffe, *E. coli*	1h
Unit of evolution	Maynard Smith, 1987	Unit which exhibits heritable variance in fitness	Horse	1a
Biological individual	Dupré & O'Malley, 2009; O'Malley, this volume	Unit of metabolic collaboration	Human-gut microbes collection; *Medicago-*Rhizobial bacteria collection	2d
Evolutionary individual	Clarke 2013; In review	Unit with capacity for heritable variance in fitness only at one level, in virtue of individuating mechanisms	Horse, Meiotic driver gene, Tasmanian devil facial cancer, aphid-*Buchnera* collection.	1i

I think there really are chunks that fit the definition underlying O'Malley's meta-bolic concept, and they really are different chunks from those picked out by my own definition. Neither will I resist the distinctiveness of the other eight concepts listed (though I might think that some definitions of them are better than others) nor, for that matter of 'My right leg and my left eyelid plus my hamster's tail.' What I discuss, instead, is the relative *usefulness* of these distinct categories and I will defend my conviction that the concept numbered '1' in my table, and espe-cially that version of it numbered '1i' stands above the other concepts in terms of usefulness: the predictive inferences it supports, the explanatory value it offers, and the range of contexts across which it offers these advantages.

These are not the only reasons we might have for maintaining a concept. Some are valuable because they capture an intuitive or historical idea, rather than for their inherent clarity or empirical applicability. There is always a tension between preserving the traditional meaning of a term, in order to avoid the communica-tive disruption brought about by revision, and seeking to enhance the work our language does for us by urging revisions. I suspect that the former urge pulls in favor of those organizational concepts numbered '2' above. My agenda here is unashamedly revisionist, however. While I do understand the reassurance offered by maintaining concept '2,' I also think that science has made available an enhanced concept: evolutionary theory is able to explain why our ancestors came to use concept '2,' as I'll explain in Section 8.6. In a nutshell, I accept O'Malley's claim that her metabolic individuals are distinct from my evolutionary individuals – but I deny that the concept she defines is useful enough to be worth holding onto.

First, note that the definition I advocate is not the same as the ones criticized by O'Malley – what *she* calls the evolutionary individual. O'Malley's targets cor-respond to the definitions numbered 1b and 1g in my table. What, instead, is the concept '1i' that I advocate?

8.3 Clarke's Levels-of-Selection Approach to Evolutionary Individuality

The concept I define refers to a kind, and the definition allows us to decide whether particular things belong in the kind group or not. The concept functions as a 'sortal' term, so it allows us to answer questions about *how many* members of the kind there are – to count individuals. Most concepts work by drawing distinctions, and my concept distinguishes individuals from *parts* of individuals, and from *groups* of individuals (Pepper & Herron, 2008). There are concepts which are similar but which define a biological individual in contrast to something distinct. For example, we might be concerned to distinguish a biological individual from a *non*-biological individual. Or between living and *non*-living things. Similarly, we might want to distinguish a biological individual from a biological *process* or *prop-erty*. These distinctions must be drawn by distinct concepts. Most commentators in the debate regarding biological individuality, sometimes also referred to as a debate about organismality, are concerned with the distinction between individ-uals, groups and parts, all of which may be assumed to be biological/alive, and all of which may be assumed to be objects as opposed to properties or processes.[2]

Evolutionary definitions distinguish biological individuals from biological parts and from biological groups by thinking about which things are treated as objects, rather than as parts or as groups, *by the process of evolution by natural selection*. Several subtly different evolutionary definitions have been proposed (see all concepts labelled number 1 in Table 8.1). My 'Levels-of-selection account' defines an evolutionary individual in terms of its possession of mechanisms that ground a capacity to participate in a process of evolution by natural selection.

Definition: An evolutionary individual is a collection of living parts which has some *capacity* for responding to selection at the between-collection level, *because of* the action of individuating mechanisms.

The relevant capacity is one that objects can have more or less of, and they can have it at multiple hierarchical levels.[3] *Exclusive* evolutionary individuals have the capacity at only their own hierarchical level. Simplifying a little, an evolutionary individual is the stuff that has the capacity to respond to natural selection. We add more detail to that description by looking to evolutionary theory to tell us what sorts of properties an object needs to have in order to respond to natural selection. Lewontin, building, of course, on Darwin, summarized these properties as reproducing with heritable variance in fitness (Lewontin, 1970). And we can add even more by detailing the sorts of mechanisms that can ground the manifestation of those properties[4] (Clarke, 2013). As an example, a bottleneck in the individual's life cycle can help to ground heritability across generations of those individuals, by sieving out genetic variation that could otherwise lead to lots of divergence between generations.

If we focus on an object that is too small it won't have the properties necessary for responding to selection – my concept says it's a mere part of an individual. Likewise, if we focus on an object that is too big, it won't have the properties either. Empirical investigation together with evolutionary theory can tell us, of any particular bundle of living matter, whether it does or doesn't have the properties necessary for responding to natural selection. The concept thus generates verdicts about the individuality of any lump of living matter – for example, mitochondria are not individuals, but parts; ecosystems are not individuals but groups; symbiotes in general are proper parts only to the extent that there are mechanisms guaranteeing their common response to selection along with their partners.

My evolutionary concept satisfies the desiderata of a good scientific concept. It supports inferences of a very useful kind. Plugged into the theory of natural selection so that it bears fitness, the evolutionary organism concept supports successful *quantitative* predictions (about the likely spread or decrease of a trait in the population), as well as generating *explanations* of the surprising design and diversity we observe around us (Clarke, 2012). It is very general, because it is applicable to objects in every single part of the tree of life, rather than being restricted to particular groups or particular eras (Clarke, 2013). So it is highly projectible, which means that it can support a wide array of inductive inferences, without being limited to particular times or places.

One of the primary functions of this evolutionary definition of the individual is to unify the 'organism,' which functions as the bearer of fitness in standard evolutionary models, with the major transitions literature (Margulis, 1970; Buss, 1987; Wilson & Sober, 1989; Maynard Smith & Szathmary, 1995). In particular, with those accounts of major transitions which treat a transition as an event in which a new individual is formed, at a new hierarchical level of selection, by merger of pre-existing individuals (Michod, 1999; Okasha, 2006; Bourke, 2011; West et al., 2015).[5] On this view, the unit which bears fitness and the unit which emerges during a process of transition are one and the same.

My concept doesn't mistake genes as the only source of heritability in evolution and certainly does not equate to Janzen's view (concept 1b) which delimits living things by genotype. Genetic homogeneity is not a necessary or sufficient condition of individuality on my account. Population genetics doesn't really settle problems of evolutionary individuality at all. It assumes that diploid organisms will bear two copies of each gene, but it doesn't make any further claims about what qualifies as an organism.[6] We can track the fitness of seeds, leaves, whole trees or even whole clones. We can even track the fitness of somatic cells or mitochondria if we choose to. We just won't generate accurate predictions about the future traits of such units if they are not functioning as levels-of-selection individuals.

8.4 Legume–Rhizobial Sets

In her chapter, O'Malley (this volume) highlights the ancient symbiotic relationship that takes place between between plants of the *Medicago* genus and a set of rhizobial bacteria, as a particularly strong example of a metabolic individual. These bacteria fix nitrogen within specially adapted nodules on the roots of the host plant, which in its turn provides those bacteria with carbon. Any plant may host ten or more rhizobial strains at the same time. Each partner is able to survive independently of the other – the symbiosis is not obligate – but each do much better when enjoying an association (Denison & Kiers, 2011).

An evolutionary view treats this case as one in which a multitude of distinct evolutionary individuals interact with one another. The plant is one such individual, and each bacterial cell is another. The combined set fails my definition of an evolutionary individual because there is no mechanism for transmission of phenptypes across generations. The bacteria are transferred horizontally, in that they are taken up from the environment early on during the plant's development. Because the bacteria spread through the soil, while the plant seeds travel above ground, the two lineages don't track one another – any plant tends to be infected by different strains of bacteria from those that infected its parent plants. This means that novel plant-bacterium phenotypes will rarely be passed on. Bacterial mutations are passed on to offspring bacteria and might do well if they are successful across a range of different plant hosts. And plant mutations are passed on to offspring plants, where they can do well if they are successful in interaction with a range of different rhizobial strains. But novel traits whose success depends on a particular bacterium-plant pair will be lost. The evolutionary fates of the different partners thus fail to coincide – they do not respond to selection as a unit.

The separation of the distinct evolutionary individuals is essential to explaining various phenomena associated with this symbiosis. For example, horizontally transmitted symbiotes tend not to co-speciate with their partner, or undergo genome reduction, because selection acts in the normal way on each separate species population. The bacteria spend part of their life cycle surviving without a plant's help, and they have to compete with one another for access to the plant hosts. This process sieves out those deleterious mutations that can accumulate in the genomes of vertically transmitted symbiotes (Bright & Bulgheresi, 2010).

If I resist the reification of metabolic individuals do I lack resources to explain 'how the plant gets by in the world' (O'Malley, this volume)? Not at all. An evolutionary approach recognizes the bacteria as a very important part of the plant's environment, and vice versa, aiming to elucidate the fitness gains each partner gets from the interaction.

One particularly strong example of an evolutionary explanation of these symbioses draws on market theory in economics (Werner et al., 2014). This view makes sense of the enduring relationship at the level of the two-species – and the footprints that long relationship has left on the genomes and phenotypes of each species – by taking account of the fact that the metabolic relationship is not fixed across the lifetimes of the partners, but is instead dynamic. The plant is able to modify the quantity of the metabolic resource it provides to the bacteria. This enables it to impose sanctions – providing less carbon and restricting oxygen supply – on those nodules which underperform their nitrogen-fixing duties (Kiers et al., 2003), or to preferentially allocate resources to top-performing nodules. This is significant in so far as it helps us to explain how the interaction remains stable against cheaters – the plant can simply withdraw its cooperation from partners that don't pay their way, and so maximize the amount of nitrogen it gets. In fact, some parasitic rhizobial strains are able to cheat this system, by sharing a nodule with a mutualistic strain. Denison & Kiers think the symbiosis survives despite this threat *because* the expected fitness benefits to mutualistic strains depend on their abundance in the local area relative to host plants (Denison & Kiers, 2004a). The theory of biological markets predicts that the 'price' demanded for any service will fluctuate according to various factors including the number of hosts as well as competition for hosts, in addition to environmental conditions such as background levels of nitrogen. Note that these models don't treat the markets as having fixed boundaries. It is assumed that the number of players in a given market will be variable, and that buyers/sellers will come and go over time (Wyatt et al., 2014).

8.5 "Nothing in Biology Makes Sense Except in the Light of Evolution"? (Dobzhansky, 1973)

There are some old reasons to think that evolutionary explanations take a very general sort of priority over others, as Dobzhansky insisted. For example, many authors have held that ultimate (evolutionary) explanations, which explain a trait by invoking a history of selective success, have priority over proximate explanations, which explain a trait by detailing its more immediate developmental or mechanical causes: *why* is always more important than *how*. However, recent authors have argued convincingly that this distinction between proximate and ultimate causes is, in the end, unsustainable (Laland et al., 2011; Calcott, 2013; Laland et al., 2013).

Real evolutionary processes involve ongoing reciprocal causation between populational and mechanical phenomena, such that the status of any particular cause as either proximate or ultimate is perspective-dependent. A metabolic phenomenon occurring as a consequence of mutation or environmental shift may be treated as proximate, and a mechanistic explanation sought for its consequences. But among these consequences may be subsequent selection driven by an effect on fitness, putting the very same phenomenon into an ultimate explanatory role. So complete evolutionary explanations will include many and varied how and why components. In explaining the emergence of a new evolutionary individual during a transition event there are numerous explananda. At each stage, we are interested in accounting for *why* the new individual emerged – ie what was the selective benefit (Calcott, 2008)? We want to know *how* it emerged – what was the series of mutations of developmental steps or environmental changes that allowed the benefit to be enjoyed? And we also want to answer a maintenance question – what prevents the interaction from breaking down? Little is served by competing these components against one another.

Brandon and Rosenberg advanced a different argument, that biological kinds have to be understood evolutionarily, because they are inherently functional. "To call something a wing, a feather, a tissue, a cell, an organelle, a gene, is at least implicitly to describe it in terms of function, ie the purpose it serves in the behavioural economy of some larger system." (Brandon & Rosenberg, 2003, p. 148) Functional definitions only make sense when justified evolutionarily, by providing a causal-historical explanation in terms of selection *for* functions. So evolution provides *constitutively causal* explanations of biological objects. A definition of an ice crystal need not invoke any historical process of formation. But in defining biological objects, to say what they are we need to say what they do ... what they are *for*. To put the point another way, biological stuff has a history, and that history leaves marks that would otherwise be inscrutable. Just as you cannot hope to understand the political affairs of a sovereign state without knowing its history, so biological objects won't make sense unless you know something of the story about how they came to be.

This case can be overstated, since there are surely large areas of biological science, medicine and physiology for example, which have been little changed since the advent of evolutionary theory. A neat way to bring out the relevant point is to ask 'What difference would it make if creationism were true?' –i.e. if it turned out that there had never been any evolution by natural selection. Some fields, such as biochemistry, might proceed largely as before, because they deal in current functions rather than selected effects. Yet there is reason to think that organisms might be a special case. There are certain features of living things that would be inexplicable, if it were the case that creationism were true.[7] One is the hierarchical compositionality that is apparent in many life forms. Why would life have a nested structure, with organelles inside cells inside multicellular agglomerations, if it had not emerged out of a piecemeal process which built new organisms out of interactions between existing ones? And, perhaps even more pressingly, why should organisms so often suffer maladies and maladaptations because their parts are struggling to get on or to reconcile their interests? That living things are compositional, and that they exhaust so many resources on complex mechanisms of conflict resolution, demand an evolutionary explanation, for they are stark relics of a particular contingent history.

This is a broad use of the term 'evolutionary' to delimit explanations. I don't limit the category to genetic or population-genetic explanations. Nor do I consider evolutionary explanations to be necessarily focused on optimality. It is important not to assume that all traits are adaptive. Some 'why' questions have developmental answers. Some traits are mere side-effects of selection for other properties. But one question for which no developmental or physical explanation will do is 'Why are there organisms?' In other words, why does life come in the form of cohesive packages of different sizes, rather than being some sort of homogeneous soup of living matter? The right answer is evolutionary. Life is packaged up so that it can participate in selection processes at various hierarchical scales.

8.6 In What Sense Does a Scientific Concept Have Priority Over Its Traditional Counterpart?

I don't want to insist on priority in a reductive sense: I don't think we should junk concept '2' and only use concept '1i' from now on, any more than we should replace Eddington's commonplace table with the scientific table (Eddington, 1927). What sort of priority do I mean then?

Suppose we consider an organizational definition, which defines the kind 'organism' as containing only those objects which are organized in a special kind of way. There is no mention of 'evolution' in this definition. And yet the organizational definition and the evolutionary definition will pick out roughly similar things – me, you, my dog, that bird. Why should this be? The organizational concept is designed to explain the features of a class of things that is in some sense obvious to human observers. They are things with which we interact, things which have agency. Their parts are connected to each other in obvious ways, so that they die if we sever the connections, for example. We have always noticed these things because it has been useful to us to notice them, much more useful than if we operated with a concept that apprehended an approaching tiger as an unconnected collection of tiger parts. We have used the concept 'organism' since long long before anybody started to formulate a theory of natural selection, it remains unnecessary to understand the theory of natural selection in order to be fairly competent with the use of the kind 'organism.' So what can the two definitions possibly have to do with one another? Why do they often pick out the same things?

The answer is that the theory of evolution by natural selection explains why there are organisms, in other words, why our world is populated with bits of stuff that act in unity, that jump out at us and succumb in their totality to our attacks. Evolution – the *process*, not the theory – created those things that are so numerous and so significant in the human psyche that we made up a special group name to describe them all – organisms. And consideration of the *theory* of natural selection – the theory that explains how evolution works – can tell us things about organisms that cannot be deduced by looking at the examples around us. It can tell us what sorts of properties organisms cannot have, for example – about impossible organisms. It can tell us about organisms that are possible but non-existent. It can tell us how organisms came into existence, and why. And we can use these sophistications to help us to correct our intuitive, folk concept. To settle cases in respect of which our intuitions are silent. And to help us understand which of the 'other' definitions of 'organism' are good ones. In this sense, I think that concept 1i is able to capture what concept 2 – the notion of one single life, one animal or plant – really is.

8.7 Worries about a 'Metabolic Individual'

O'Malley advocates for a concept of the biological individual that is defined in terms of metabolism – the 'biologically structured conversion of energy and carbon' (O'Malley & Powell, 2015). A metabolic individual is a collection of diverse parts, which are engaged in metabolic collaboration – energy exchange – with one another. They range from 'electron trades within and between cells' to *Medicago*-Rhizobia sets up to 'global biogeochemical cycles' (ibid.). I assume, given that viruses are meant to be excluded from the class of metabolic individuals, that some threshold of metabolic autonomy must also be met (O'Malley, this volume). The parts must exchange energy with one another in such a way that the sum is self-sustaining, to some extent, I think.

Metabolic interactions, I assume, are fairly ubiquitous. All living things rely, to a greater or lesser extent, on products or processes performed by other living things, in order to survive and reproduce. For example, humans rely on a mate in order to produce offspring, and they rely on a variety of gut flora to allow them to digest their food and maintain their immune systems. Furthermore, modern humans rely on their families, communities and societies more generally in order to meet their needs. They rely on plants and bacteria to oxygenate the atmosphere so that they can breathe. They rely on various other life forms to act as food sources. Humans exchange energy or carbon with all these other living things. So I wonder about the threshold amount or type of interaction, or sense of autonomy, that would circumscribe a human as a metabolic unit. If anything upon which a living thing depends for survival qualifies as a metabolic part then it seems natural to think that there can only be one metabolic individual in existence – the whole planet.

Even accepting, as I do, a significant, perhaps even starring, role for metabolism as explanatory of evolutionary individuals, is this explanatory role best served by a concept of a metabolic *individual*? Individuals, in whatever domain they occur, are particulars. Unlike classes, particulars are supposed to have spatio-temporal identity. They have births and deaths, rather than existing timelessly. We can expect answers to questions about whether one individual is the same numerical individual as another or not, about what sorts of events the individual cannot survive, about what things are their parts.

Some metabolic interactions will occur within a community in which different members constantly come and go. In a typical ecosystem, or a microbial community, any particular organism or cell may participate for some period of time and then leave. The legume-rhizobia collaboration may be more stable, in so far as one plant might retain the same ten or so strains of bacteria in its root nodules throughout its life. Nonetheless, there will rapid turnover of particular bacteria cells during that time. Furthermore, the boundaries of the set that actively engages in metabolic interaction will vary over time, as different nodules vary their level of cooperation or of sanction.

Where does the metabolic individual begin and end, in these cases? Does any metabolic individual persist across changes to the identity of its parts? Can it survive the death of a single participant? What if the parts remain but temporarily suspend their active metabolic engagement? Is the plant, but no particular bacterium, a necessary participant in the interaction, and if so, why? Maybe the individual persists so long as a threshold energy transfer is maintained, regardless of the identity of the separate collaborators?

If we are going to make room, in our ontology, for O'Malley's metabolic individual concept, then these questions need to be answered, and the answer needs to be motivated – there needs to be some explanation or generalization that depends upon the question being answered that way, and not some other way. I suspect that a greater degree of light would be shed, instead, by thinking about metabolism as an extremely important *process* in which biological individuals engage.

A view which treats a whole legume–rhizobia set as a cohesive individual overlooks any conflict between the parts, and the variation across different nodules in levels of metabolic interaction. A view which focuses on the category of metabolic exchange without attention to the identities of the exchangers doesn't take account of the effects of infection by multiple competing strains of bacteria. A metabolic view also obscures the fact that many hosts will interact with multiple mutualists simultaneously, rather than with only one. For example, plants of the genus *Medicago* engage in collaboration with mycorrhizal fungi as well as with rhizobial bacteria. The fungi thus ends up indirectly interacting with the bacteria, in so far as each relies on the other to sustain its partner. This can drive further interesting dynamics such as divisions of labor between different mutualists (Werner et al., forthcoming). These dynamics are not captured by a view which cuts things up according to the particular metabolites transferred.

8.8 Conclusions

I have tried to argue that my concept of the evolutionary individual is more useful than O'Malley's concept of the metabolic individual. For example, the evolutionary concept is equipped to meet all the explanatory challenges accomplished by the metabolic view of the legume–rhizobia collaboration, and it supports several further explanations of phenomena about which the metabolic view can say little. Even if there are particular contexts in which a different view of the individual is most useful, we might say that the concept of the evolutionary individual is useful across the biggest range of explanatory contexts.

I don't deny the significance, perhaps the pivotal, significance of metabolism to making evolutionary individuals what they are. I agree that metabolic perspectives can contribute 'additional explanatory resources' for evolutionary theory. Yet, I do aim to cast doubt on the usefulness of focusing particular attention on units delimited by metabolism. I suggest that the explanatory value of thinking about metabolic processes is not best served by postulating metabolic *individuals*. The metabolic interactions, such as those occurring between trees and their mycorrhizal fungi, are fascinating and important. But creating a concept to unify the partners in such an interaction is less useful than, for example, applying biological market theory to explicate the evolutionary forces that sculpt those interactions.

Notes

1 Although whether all these concepts do in fact meet those standards is open to debate.
2 One implication of this is that my account doesn't deal with viruses. Only *if* we decide to consider them as alive, then I say they are evolutionary individuals.
3 Note the implication of this that some thing may be an individual as well as being a part, or a group, if the relevant capacity is grounded at different hierarchical levels to intermediate degrees.
4 There are two types of mechanisms which are together sufficient (a) policing mechanisms which prevent the object's parts from undergoing differential selection and (b) demarcation mechanisms which enable to object to compete against others of its type. There are many different ways to realise these mechanism types, and they vary across different lineages. See Clarke (2013) for details.
5 Note that this is not the way O'Malley understands major transitions.
6 Genes themselves are not evolutionary individuals, on my view, except on those occasions when they are selected separately from the rest of the genome, as in the case of meiotic driver genes.
7 Or would at best require ad hoc explanations.

References

Borges, Jorge Luis. (1966). The analytical language of John Wilkins. In Ruth L. C. Simms (Trans.), *Other inquisitions, 1937–1952* (pp. 101–105). Austin, TX: University of Texas Press.

Bouchard, Frédéric, & Huneman, Philippe (Eds.). (2013). *From groups to individuals: evolution and emerging individuality.* Cambridge, MA: MIT Press.

Bourke, Andrew F. G. (2011). *Principles of social evolution.* Oxford: Oxford University Press.

Brandon, Robert N. (1999). The units of selection revisited: the modules of selection. *Biology and Philosophy, 14*(2), 167–180.

Brandon, Robert, & Rosenberg, Alexander. (2003). Philosophy of biology. In Peter Clark & Katherine Hawley (Eds.), *Philosophy of science today* (pp. 147–180). Oxford: Oxford University Press.

Bright, Monika, & Bulgheresi, Silvia. (2010). A complex journey: transmission of microbial symbionts. *Nature Reviews Microbiology, 8*(3), 218–230.

Buss, Leo W. (1987). *The Evolution of individuality.* Princeton, NJ: Princeton University Press.

Calcott, Brett. (2013). Why the proximate–ultimate distinction in misleading, and why it matters for understanding the evolution of cooperation. In Kim Sterelny, Richard Joyce, Brett Calcott, et al. (Eds.), *Cooperation and its evolution* (pp. 249–264). Cambridge, MA: MIT Press.

Clarke, Ellen. (2012). Plant individuality: a solution to the demographer's dilemma. *Biology & Philosophy, 27*(3), 321–361.

Clarke, Ellen. (2013). The multiple realizability of biological individuals. *The Journal of Philosophy, 110*(8), 413–435.

Denison, R. Ford, & Kiers, E. Toby. (2004a). Lifestyle alternatives for rhizobia: mutualism, parasitism, and forgoing symbiosis. *FEMS Microbiology Letters, 237*(2), 187–193.

Denison, R. Ford, & Kiers, E. Toby. (2004b). Why are most rhizobia beneficial to their plant hosts, rather than parasitic? *Microbes and Infection, 6*(13), 1235–1239.

Denison, R. Ford, & Kiers, E. Toby. (2011). Life histories of symbiotic rhizobia and mycorrhizal fungi. *Current Biology, 21*(18), R775–R785.

Dobzhansky, Theodosius. (1973). Nothing in biology makes any sense except in the light of evolution. *American Biology Teacher, 35*, 129–129.

Dupré, John. (1993). *The disorder of things: metaphysical foundations of the disunity of science.* Cambridge, MA: Harvard University Press.

Dupré, John, & O'Malley, Maureen A. (2009). Varieties of living things: life at the intersection of lineage and metabolism. *Biology & Philosophy, 1.*

Eddington, Arthur S. (2012). *The nature of the physical world: Gifford lectures (1927).* Cambridge: Cambridge University Press.

Ereshefsky, Marc, & Pedroso, Makmiller. (2015). Rethinking evolutionary individuality. *PNAS USA, 112*(33), 10126–10132.

Folse, Henri J., III, & Roughgarden, Joan. (2010). What is an individual organism? A multi-level selection perspective. *The Quarterly Review of Biology, 85*(4), 447–472.

Frank, Steven A. (2003). Repression of competition and the evolution of cooperation. *Evolution, 57*(4), 693–705.

Gardner, Andy, & Grafen, Alan. (2009). Capturing the superorganism: a formal theory of group adaptation. *Journal of Evolutionary Biology, 22*(4), 659–671.

Godfrey-Smith, Peter. (2009). *Darwinian populations and natural selection.* Oxford: Oxford University Press.

Godfrey-Smith, Peter. (2013). Darwinian individuals. In Peter Godfrey-Smith (Ed.), *From groups to individuals: evolution and emerging individuality* (pp. 17–36). Cambridge, MA: MIT Press.

Gould, Stephen Jay, & Lloyd, Elisabeth A. (1999). Individuality and adaptation across levels of selection: how shall we name and generalize the unit of Darwinism? *Proceedings of the National Academy of Sciences, 96*(21), 11904–11909.

Griesemer, James. (2000). Development, culture, and the units of inheritance. *Philosophy of Science, 67*, S348–S368.

Janzen, Daniel H. (1977). What are dandelions and aphids? *The American Naturalist, 111*(979): 586–589.

Kant, Immanuel. (2007). *Critique of judgment.* (Nicholas Walker, Ed., James Creed Meredith, Trans.). Oxford: Oxford University Press. [Originally published in 1790]

Kiers, E. Toby, Rousseau, Robert A., West, Stuart A., et al. (2003). Host sanctions and the legume–rhizobium mutualism. *Nature, 425*(6953), 79–81.

Laland, Kevin N. (2013). More on how and why: a response to commentaries. *Biology & Philosophy, 28*(5), 793–810.

Laland, Kevin N., Sterelny, Kim, Odling-Smee, John, et al. (2011). Cause and effect in biology revisited: is Mayr's proximate-ultimate dichotomy still useful? *Science, 334*(6062), 1512–1516.

Lewontin, Richard C. (1970). The units of selection. *Annual Review of Ecology and Systematics, 1*(1), 1–18.

Lloyd, Elisabeth. (2005). Units and levels of selection. In Edward N. Zalta (Ed.), *The Stanford Encyclopedia of Philosophy (Fall 2005 Edition).* URL = <http://plato.stanford.edu/archives/fall2005/entries/selection-units/>

Margulis, Lynn. (1970). *Origin of eukaryotic cells: evidence and research implications for a theory of the origin and evolution of microbial, plant and animal cells on the precambrian Earth.* New Haven: Yale University Press.

Maynard Smith, John. (1987). Evolutionary progress and levels of selection. In John Dupré (Ed.), *The latest on the best: essays on evolution and optimality.* Cambridge, MA: MIT Press.

Maynard Smith, John, & Szathmáry, Eörs. (1995). *The major transitions in evolution.* New York: Freeman.

Michod, Richard E. (2000). *Darwinian dynamics: evolutionary transitions in fitness and individuality.* Princeton, NJ: Princeton University Press.

Michod, Richard E. (2005). On the transfer of fitness from the cell to the multicellular organism. *Biology & Philosophy, 20*(5), 967–987.

O'Malley, Maureen A., & Powell, Russell. (2016). Major problems in evolutionary transitions: how a metabolic perspective can enrich our understanding of macroevolution. *Biology & Philosophy, 31*(2), 159–189.

Okasha, Samir. (2006). *Evolution and the levels of selection.* Oxford: Oxford University Press.

Pepper, John W., & Herron, Matthew D. (2008). Does biology need an organism concept? *Biological Reviews, 83*(4), 621–627.

Pradeu, Thomas. (2010). What is an organism? An immunological answer. *History and Philosophy of the Life Sciences, 32,* 247–267.

Queller, David C., & Strassmann, Joan E. (2009). Beyond society: the evolution of organismality. *Philosophical Transactions of the Royal Society of London B: Biological Sciences, 364*(1533), 3143–3155.

Rosenberg, Alex. (2001). How is biological explanation possible? *The British Journal for the Philosophy of Science, 52*(4), 735–760.

Sober, Elliott, & Wilson, David Sloan. (1999). *Unto others: the evolution and psychology of unselfish behavior.* Cambridge, MA: Harvard University Press.

Werner, Gijsbert D. A., Strassmann, Joan E., Ivens, Aniek B. F., et al. (2014). Evolution of microbial markets. *Proceedings of the National Academy of Sciences, 111*(4), 1237–1244.

Werner, Gijsbert D. A., Wyatt, Gregory A. K., Kiers, E. Toby, et al. (forthcoming). *Specialised mutualistic services or providing it all? The evolution of division of labour among mutualistic symbionts.*

West, Stuart A., Fisher, Roberta M., Gardner, Andy, et al. (2015). Major evolutionary transitions in individuality. *Proceedings of the National Academy of Sciences,* 201421402.

Wilson, David Sloan, & Sober, Elliott. (1989). Reviving the superorganism. *Journal of Theoretical Biology, 136*(3), 337–356.

Wilson, Jack. (1999). *Biological individuality: the identity and persistence of living entities.* Cambridge: Cambridge University Press.

Wyatt, Gregory A. K., Kiers, E. Toby, Gardner, Andy, et al. (2014). A biological market analysis of the plant-mycorrhizal symbiosis. *Evolution, 68*(9), 2603–2618.

Study Questions for Part IV

1. What is an evolutionary individual according to Clarke? Is the legume–rhizobi an evolutionary individual?
2. What is a metabolic individual according to O'Malley? Is the legume–rhizobi a metabolic individual?
3. According to O'Malley, what does it mean to explain individuals by their validity? And what does it mean to explain them by their fecundity?
4. According to O'Malley, what explanatory advantages are there to distinguishing between biological and evolutionary individuals?
5. What are some reasons to favor Clarke's evolutionary definition over O'Malley's metabolic definition as a definition of a biological individual?

Part V

Is Chance Ontologically Fundamental?

9 Chance and the Great Divide

Ned Hall

9.1 Introduction

David Lewis' classic 1980 paper "A Subjectivist's Guide to Objective Chance" helped set the terms for one of the most vibrant debates in philosophy of science and metaphysics of the last 35 years. Lewis saw, as had others before him (e.g. Carnap, 1945), that probability theory has distinct applications to the epistemic and physical domains: very roughly, we must distinguish probability as a way of representing *rational credence* from probability as a way of describing certain aspects of the *physical world*. We therefore require more than one answer to the question "What is probability?" along with an account of how these answers relate to each other. We'll see below that Lewis set some confused terms for the debate that followed him. But the debate itself has been enormously productive. And one absolutely central point that has emerged from it concerns the significance of deep metaphysical questions about the nature of the physical modalities (laws of nature, causation, chance) for our understanding of physical probability, and its connection to rational credence. That significance is what this chapter explores. After laying out a useful framework (Section 9.2), and saying just what those deep metaphysical questions center on (Section 9.3), we'll explore in detail just how different "chance" (as we'll henceforth label the kind of probability that attaches to the physical world) looks, depending on the correct answer to those questions (Sections 9.4–9.8).

9.2 Framework

Lewis took for granted a certain understanding of the *structure* of chance, on which the canonical expression of a fact about objective chances has this form:

The objective chance, at time t, that proposition P is true, is x.

Note three happy-making features of this understanding. It is admirably clear about what chances attach to; it steers us away from the confused thought that there is such a thing as *the* chance of P; but it does so without steering us toward the confused frequentist idea that chances are had only relative to a choice of reference class. Still, it builds in a different confused thought, which is that chances are, in the first instance, defined *at times*. What of worlds such as our own, whose relativistic spacetime structures serve up no objective basis for talk of "times"? And even in worlds with Newtonian or neo-Newtonian spacetimes, we *might* want to allow that chances can attach to their initial conditions; while these might be chances *about* a time (namely, the first time), they seem not to be chances *at* that, or any other, time. Our most basic way of talking about chance should not rule out this possibility.

Here is a better – because more flexible – framework. Take each possible world w to have associated with it a single "ur-chance" function $urch^w$, defined over a set S of possible worlds (which we will take to include w).[1] For a given proposition P, $urch^w(P)$ is the chance according to w that P is true, and is simply the sum of the probabilities $urch^w$ attaches to worlds in S that make P true.[2] Chance is thus modeled as a contingent feature of a world; but it is, in the first instance, a feature of that world, not of a time within it.

The framework readily allows for chances at times, in two different ways. Given some time t, let S_{tw} be the proposition true in a world v iff the state of v at t is exactly the same as the state of w at t; let H_{tw} be the proposition true in a world v iff the history of v up through t is exactly the same as the history of w up through t. Then either $urch_w(\bullet \mid S_{tw})$ or $urch_w(\bullet \mid H_{tw})$ can serve as "the" t–chances (in w). (Which one should we choose? We can, and should, let the answer depend on what particular use we have for a notion of chances-at-times.)

The framework also allows for chances over initial conditions – even when there is no first moment of time – as follows: Say that two worlds w and v "start out the same" iff there is some moment of time t such that they are exactly qualitatively alike up to t.[3] *Starting out the same* is an equivalence relation; so it partitions S.[4] We can thus take the chances over initial conditions to be given by the values $urch_w$ attaches to the elements of this partition.

I'm going to assume this framework henceforth. Note that it is quite flexible – much more flexible than you might need, depending on your views about the nature and source of objective chances. For example, if you think about chances as, in the first instance, characterizing the lawful transition from earlier to later states (see for example Maudlin, 2007; more on this conception below), you'll likely consider it incoherent to attach chances to initial conditions, and you'll likely have no interest in "backwards looking" chances (chances defined *at* times, concerning matters *prior* to those times). That doesn't prevent you from using the framework to represent what you think is real, concerning chances. It's just that surplus mathematical structure will come along for the ride, structure to which you attach no physical significance.

9.3 The Great Divide

A philosophical account of the physical modalities faces a basic choice point. On the first, "Humean" branch, laws, causes, and chances are nothing more than certain kinds of *patterns in the nonmodal phenomena*. On the second, anti-Humean branch, the physical modalities have some sort of objective reality above and beyond that of the nonmodal facts, and consist in some kind of *fundamental constraints* on how these facts unfold.

For more precision, we need an account of what "nonmodal" and "nothing more than" come to. As to the first, we might begin by focusing on the most fundamental physical magnitudes instantiated at our world, and equate the (controversial) claim that they are "nonmodal" with the thesis that the *metaphysical* possibilities concerning their instantiation coincide with the *purely combinatorial* possibilities. (For discussion, see Hall, 2010.) As to "nothing more than," a good start is contained in the well-known thesis of "Humean supervenience" (see, e.g., Lewis, 1994): no two possible worlds differ in their physical modalities without differing, somehow, in their nonmodal facts.[5] Thus, an anti-Humean account can earn that status either by denying that fundamental physical magnitudes are nonmodal,[6] or by positing some additional ingredient to reality from which the physical modalities derive.[7]

Like Jenann Ismael (this volume), I think that a deeper issue lurks behind this divide, one concerning the nature of explanation and understanding.

Begin with a truism: scientific inquiry aims to provide us with *understanding* of the world. Here is a fundamental philosophical question about such understanding: Does it consist in the possession of a *special kind of information*, or does it rather consist in having one's information *organized in a special sort of way*? ("Some of both" is also an option!)

It will help to have some illustrations. For the "special kind of information" position, examples seem awfully easy to come by. The window broke. Why? What *explains* the breaking? What do we need to know, in order to *understand* how this came about? Just this: Suzy threw a rock at it, which struck the window with sufficient force to break it. Of course there's plenty more we could add, in order to enrich our understanding; but it is striking that understanding seems to *begin*, at least, with knowledge of the breaking's *causes*. Which is to say, knowledge of a special kind of information about the breaking.

A second example. As a planet orbits the sun, the line joining it to the sun sweeps out (to a very close approximation) equal areas in equal times; this is Kepler's second law. What explains this regularity? Newton's laws of motion, together with the fact that the gravitational force of the sun on a given planet dominates all other forces on that planet, *and* is a central force (i.e., a force acting on a line joining the sun with the planet).[8]

Now for a very different example. Consider the following initial segment of an infinite sequence of natural numbers:

1,1,1,2,3,2,1,3,5,4,2,5,7,8,3,7,9,16,5,11,11,32,8,13,13,64,13,17,...

Perhaps you've figured out the rule that generates the sequence. Perhaps, on the other hand, you find it confusing. You don't understand it. You don't know why it has the form it does. If so, the following way of reorganizing the initial segment will make things clear:

1,	1,	1,	2,
3,	2,	1,	3,
5,	4,	2,	5,
7,	8,	3,	7,
9,	16,	5,	11,
11,	32,	8,	13,
13,	64,	13,	17,...

Looking down the columns, we see that the sequence is just an interleaving of the odd numbers, powers of 2, fibonacci numbers, and prime numbers. Once you see this, you understand the sequence. But not by acquiring a special sort of information about it. (The sequence is, after all, not the sort of thing that has "causes," or that "metaphysically depends" on anything else.) To me, examples like this evoke in its purest form the idea that to understand some subject matter is to organize one's information about it in the right sort of way. (Of course, actual explanatory practice in mathematics is the best place to look, for more serious examples.)

Here is the key upshot. One broad approach sees an explanation of some phenomenon as consisting in information about *objective constraints that apply to it* (that's what's "special"): thus, an explanation of an event might detail its *causes*; an explanation of some unbroken regularity might cite the *laws* that issue in it; an explanation of some robust statistical patterns might appeal to underlying *chances* that generate them. In each of these cases, the conceptual distance between what is to be explained and what does the explaining argues in favor of (without outright implying) an *anti-Humean* account of the physical modalities: the core idea is that since these modalities are, or reflect, explanatory constraints on the (nonmodal) phenomena, they cannot be identified with, or even reduce to, such phenomena.[9] But another broad approach see an explanation of some phenomenon as succeeding to the extent that it *fits that phenomenon into some suitably broad, organizing framework*; following one of the few worked-out philosophical accounts of explanation along these lines (Kitcher 1989), we might call this "unificationism" about explanation. What makes such a framework *succeed*, when it comes to conferring understanding, is not that it contains information about "underlying constraints," but that it bears the right sort of logical or more broadly epistemic relationship to the particular phenomena we wish to understand. And that is an approach that argues in favor of (again, without outright implying) a *Humean* account of the physical modalities.

9.4 Two Concepts of Chance

Zeroing in on the concept of "chance," now, we should appreciate that, on a very natural (if not inevitable) way of developing Humeanism and anti-Humeanism, each proceeds from a profoundly different understanding of the *point* of such a concept. For the Humean, it is a concept that plays a role in the construction of a framework particularly suited to *organizing statistical information*, to *locating* particular statistical phenomena within some more comprehensive scheme. For the anti-Humean, it is a concept of a metaphysically distinctive kind of *partial causal-explanatory constraint*, one that must be invoked in order to properly understand (among other things) how statistical phenomena *are generated*. This stark difference in orientation has, as we're about to see, a profound effect on the shape of Humean vs anti-Humean accounts of chance.

We'll need a working example. To find real-world examples of physical theories that posit objective chances, we could turn to quantum mechanics – in particular, versions of quantum mechanics that treat the "collapse of the wave function" as an objective physical phenomenon.[10] But these examples require quite a bit of setup, so I am going to make do with a toy example instead. Consider a world consisting of particles that sometimes enter what we'll call "T-junctions." When a particle enters such a junction, it has a certain chance r of exiting to the right, and a chance (1 − r) of exiting to the left. We should think of r as a fundamental constant that helps characterize the laws at this world.[11]

9.4.1 Anti-Humean Chances

Anti-Humeanism is a big tent, and it won't profit us to survey the full variety of accounts of objective chance that it encompasses. Instead, I'll highlight a particularly simple account, drawn from Maudlin (2007).[12] This account takes the description of the laws and chances of our world more or less at face value. There is a totality of facts about how the particles happen to move – in particular, which direction they happen to take out of each T-junction they enter. As an additional, nonsupervening metaphysical ingredient, there are fundamentally stochastic dynamical laws. These laws "directly assign" (Maudlin's words) *transition probabilities*: probabilities that the world will evolve in a certain way after time t, given that it is in such-and-such a state at t. As such, the laws act as a kind of global causal constraint on how the world can evolve. And that's it – that's all there is to say.

9.4.2 Humean Chances

Given that a Humean approach to physical modalities will see them as constituted, somehow, by patterns in the nonmodal phenomena, a natural approach to chances, in particular, will see them as constituted by *statistical* patterns in the nonmodal phenomena. Thus, early Humean approaches[13] saw chances as simple frequencies, probabilistic laws as nothing more than statistical (as opposed to universal) generalizations. By now, however, we have available to us much more sophisticated and nuanced Humean accounts – and this, thanks to Lewis' introduction of the "best system" account, and its further development over the last 20 years or so.[14]

To get the key idea into view, start with best system accounts of *non*-probabilistic laws. These aim to provide a method for selecting, from among all the true statements concerning the nonmodal facts, some elite set of statements that collectively deserve to be singled out as "laws." What distinguishes them is not that they most directly reflect the underlying modal constraints or physical necessities that govern our world. Instead, laws count as such thanks to some sort of *epistemic* virtue that they collectively possess. Lewis and many following him take the virtue in question to be a kind of optimal combination of simplicity and informativeness.[15]

Let me just observe that while it is certainly natural for a Humean to privilege simplicity and informativeness in this way, doing so is neither obligatory nor without potentially serious costs. (On the costs, see Hall, 2010, Hicks, 2014; on an important alternative – though one that remains firmly committed to the idea that what distinguishes laws is a certain epistemic virtue – see Hicks, 2014.) For now, though, we'll stick with the Lewisian conception of 'bestness.'

How shall we extend the best system account so as to accommodate probabilistic laws? Return to our example of the T-junction world. If Humeanism is correct, then all there is to this world, fundamentally, is a totality of (nonmodal) facts about particle motions, and in particular facts about the directions particles take out of the T-junctions they enter. (So: no facts about *chances*, that a candidate system might need to capture.) What must a candidate system do, to achieve an epistemically optimal summary of these facts? It can keep things simple but uninformative: "Whenever a particle enters a T-junction, it exits either to the left or to the right." It can pack in a lot of information – but, since there are no simple-to-state regularities that distinguish left-exits from right-exits, such information will cost a lot in simplicity. It can occupy a middle ground: "61.8% of exits are left-exits." But though simple, that's much less informative than it might seem, as it implies nothing about *correlations* (e.g., what percentage of left-exits are followed by right-exits?).

What to do? This: Let a candidate system consist of two parts. The first part is just a set of true sentences. The second is a measure – one with the mathematical structure of a *probability* measure[16] – over the set of worlds compatible with these sentences. If a candidate system comes out "best," then the worlds compatible with *its* sentences thereby qualify as the "nomological possibilities," and *its* measure counts as encoding the "ur-chances." Fine; but what exactly does "bestness" come to, for systems with this extra probabilistic ingredient? Lewis added a third desideratum: That system is best which optimally balances simplicity (of both its set of sentences and its measure), informativeness (of its set of sentences), *and* the extent to which its measure "fits" the actual world – namely, by assigning that world a high value.[17] But a better idea, I think, is to stick with the twin criteria of simplicity and informativeness, and treat a measure as "informative," to the extent that it could be successfully used as a basis for prediction. (For a defense of something close to this idea, see Loewer, 2004; Albert, 2015b.) I'll simply assume that this can be done – and that ur-chances, so understood, will typically bear some close relation to frequencies. (Thus, we should expect the ur-chances for our T-junction world to assign exit probabilities equal to the exit frequencies.)[18]

So we have a sharp contrast between, on the one hand, a conception of chances as reflecting partial causal constraints that the fundamental laws place on how the world may evolve, and, on the other hand, a conception of chances as merely summarizing the large-scale statistical structure of the world. Let's take a look at how this contrast plays out with respect to several two key topics: the univocality of chance, and the connection between chance and rational credence.

9.5 Imperialism

Consider our T-junction world. Suppose that the fundamental dynamical laws assign a chance of left-exit of 0.9. Suppose these laws treat each exit event as independent of all others. Then consider the following structure, which we'll call a "Maze":

When a particle enters a Maze (at dot 1), it first encounters the central T-junction (dot 2); it can either (i) go right to the T-junction at dot 3, then right again, in which case it returns to the central T-junction; (ii) go left to the T-junction at dot 4, then left again, in which case it returns to the central T-junction; (iii) go

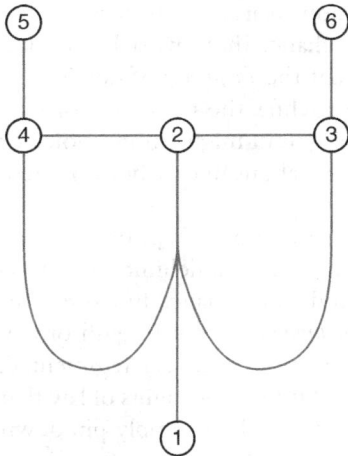

Figure 9.1

left then right, in which case it exits at dot 5; (iv) go right then left, in which case it exits at dot 6. Suppose each of these outcomes takes exactly 1 second to unfold. Then observe that, once the laws governing behavior at T-junctions are settled, *so too* are the chances pertaining to Maze behavior. If a particle enters a maze at time t, then its chance of having exited by t + n seconds is exactly $1 - (0.82)^n$; and its chance of exiting at dot 5 is exactly 0.5. This case illustrates a general point: the chances pertaining to fundamental events *settle*, in imperialistic fashion, the chances of any nonfundamental events. (Here, an exit from a T-junction counts as a fundamental event, an exit from a Maze as nonfundamental.)

Or at least, all that is so, on an *anti-Humean* conception of chance. But once again, matters look different, on the rival approach. As a warm-up, suppose we adopted a simple frequentist approach to chance, according to which the chance that a particle left-exits a T-junction is *identified* with the frequency with which particles left-exit T-junctions, and the chance that a particle dot-5-exits a Maze is likewise *identified* with the frequency with which particles dot-5-exit Mazes. Then these two chances will be almost *completely independent of each other*. Almost, because in the limiting case where the frequency of left-exits is 0 or 1, particles that enter mazes won't ever leave. But for any other left-exit frequency f_1, and any desired dot-5-exit frequency f_2, we can always design the history of particle behavior so that both f_1 and f_2 obtain.[19] Simple frequentism, then, leads directly to a denial of imperialism.

Fine; but what about a properly sophisticated Humean approach? At the very least, imperialism looks as though it is entirely optional, on the Humean approach (pending an issue to be addressed shortly). You can build it in easily enough, simply by stipulating that candidate systems include at most one ur-chance function. But you may not have to build it in – and in some circumstances, there may be powerful reason not to.

As an illustration, consider a T-junction world populated by a gazillion Mazes. If imperialism holds, and the one ur-chance function treats exit chances as constant and independent, then it's easy to see that the chance that a particle entering a Maze exits at dot 5 is exactly 0.5. So suppose that the *frequency* of dot-5-exits is quite a bit different from this — say, 0.278. Meanwhile, the frequency of left-exits is (say) exactly 0.5. Consider how a candidate system might, in its choice of ur-chance function *or functions*, aim to be informative about one or both of these statistical patterns.

At this point, we must pause over an issue that we've mostly kept off stage, up to now. A candidate system specifies both a set of sentences that delimit the nomological possibilities, relative to the given world w, and an ur-chance function over these possibilities. These sentences belong to some *language* — not English or any other natural language, surely, but something much more precisely regimented. And it is by now a commonplace in the literature on Humean accounts of law that there must be *restrictions* on this language, else the set can oh-so-simply pin down w exactly, by including just the sentence "everything is F," where the (primitive) predicate "F" is understood to express a property had by a thing x in a world v exactly if v = w. In the same way, the candidate system must also draw on some language in order to specify its ur-chance function; and for the same reason, there must be some restrictions on *this* language as well.

Perhaps these restrictions are so severe that, in the case at hand, only *one* language will meet them: namely, a language whose primitive vocabulary can express such properties as "being a T-junction" (along with whatever other properties and relations characterize the constituents of this world at the most micro-physical level), but *not* such properties as "being a Maze."[20] If so, then imperialism (as applied to this world, anyway) will likely follow. For either the properties of being a Maze, exiting at dot 5, etc. will be definable (in the one acceptable language) in a relatively *simple* manner, or not.[21] If the former, then a system may score better on informativeness by specifying an ur-chance function that says (in translation!) that the chance that a particle entering a Maze exits at dot 5 is 0.278. (Though note that this will still entail a good deal of added complexity: for example, this ur-chance function cannot treat particles' behavior at T-junctions as probabilistically independent.) If the latter, saying this would cost too much in simplicity, and so the winning system must rest content with an ur-chance function that sets the chance of left-exiting at 0.5, treats these chances as independent, and therefore *also* implies (albeit implicitly) that the chance that a particle entering a Maze exits at dot 5 is exactly 0.5.

But suppose the restrictions aren't so severe. Suppose, in fact, that it is perfectly permissible for the content of an ur-chance function to be specified directly in terms of the chances it assigns to Maze behavior — and this, even though (say) expressing "is a Maze" in the language of fundamental physics is no simpler than expressing "is a human." That is, a candidate system may make use of (at least) two distinct languages, one of which has among its primitive vocabulary such predicates as "is a T-junction" and "exits left," and the other of which has among its primitive vocabulary such predicates as "is a Maze" and "exits at dot 5." And it may thereby earn extra points for informativeness, by including one ur-chance function that conveys a great deal of information about the statistics of T-junction exits, and *another* that conveys a great deal of information about the statistics of Maze exits.

Of course it will be tempting to ask, "okay, but which ones are the *real* chances?" That temptation just reflects the deep intuitive pull of imperialism. But for a Humean, the right answer is that, in such a situation, there just is no such thing as "the" chances. It's not that there is something *indeterminate* about the chances, in the world we are imagining. No; there are, perfectly determinately, two chances, for any given event.[22]

Worlds like the one we are now considering – worlds whose statistics support, given Humeanism, anti-imperialist chances – are much more than a mere curiosity. For they reveal an added and fairly serious complication that will confront any attempt to explain the connection between Humean chances and rational credence. The next section explores this connection.

9.6 The Status of Credence–Chance Principles

Lewis' "Subjectivist's Guide" (1980) made a persuasive case that a rational agent's credence in a given proposition should bear a certain connection to her credence concerning that proposition's chances of coming true. Unfortunately, Lewis added an unnecessary complication, insisting that a proper statement of this connection needed to make use of a notion of "admissibility": Lewis thought, roughly, that one's credence in a proposition P ought to reflect one's credence concerning P's chances, *provided* that one possessed no evidence that was relevant to P not merely by being relevant to its chances. But if chance merits this kind of epistemic respect because it is, in effect, a certain kind of expert, then the right thing to do is to simply incorporate any "inadmissible" information into the statement of our credence-chance principle. Easily done. Within our ur-chance framework, an exact expression of Lewis' key idea is simple. Let Cr be some rational "initial" credence: a function that represents one possible way that perfectly rational opinion could be, prior to the incorporation of any evidence. Let A and E be any two propositions. Let Pr be some probability function, and let X be the proposition true at a world w iff w's ur-chance function is Pr. Then

$$Cr(A \mid E \,\&\, X) = \Pr(A \mid E \,\&\, X). \qquad (\star)$$

(See Hall 2004 for more discussion and defense of this principle.)

Shortly, we'll see one reason why, if Humeanism is correct, this principle won't do, after all (a reason you may have already spotted, given the foregoing discussion of imperialism). We'll also see that an adequate replacement may need to reintroduce the notion of admissibility. But before getting to all that, let's pause to unpack the principle a bit.

It may seem far from obvious how to apply this principle in practice. Happily, given certain background assumptions, it will imply something more user-friendly:

$$Cr\left(A \mid E \,\&\, ch_t\left(A\right) = x\right) = x, \qquad\qquad (\star\star)$$

where "$ch_t(A) = x$" says that the time t chance of A is x, and E is entirely about history prior to t.[23]

What should we make of principle $(\star)$? Well, it certainly seems that there must be *some* intimate epistemic connection between credence and chance, and $(\star)$ unquestionably serves as an elegant and plausible option for what this connection might be. But Lewis claimed much more for it (or rather, for the version he preferred, which incorporated his notion of "admissibility"). To begin, he claimed that this principle "captures all we know about chance" (1980, p. 86) – very likely a mistake, and at any rate not the sort of thing one ought to just baldly assert, without argument.[24] Just as confused, I think, was his insistence that anti-Humean accounts face an insuperable obstacle in their inability to justify $(\star)$. (See especially his (1994).) For unless we wish to embrace an utterly insane kind of inductive skepticism, we need to recognize that rational opinion must be constrained by substantial a priori commitments concerning what the world is like, in contingent respects. If anti-Humeanism is correct, the content of some of these basic epistemic constraints will be given directly in terms of the metaphysical structure that distinguishes anti-Humeanism from Humeanism. And we have, as yet, no good reason for denying that principle $(\star)$ *just is* one of these basic constraints. Now advances in epistemology might uncover such reasons – reasons, that is, for thinking that $(\star)$ or something like it can only be acceptable as a *derivative* principle of epistemology. But unless and until they do, it is pointlessly obstructionist to pretend that the very tenability of anti-Humeanism hinges on whether it can provide some independent justification of the credence-chance connection.[25]

Matters will be different, should Humeanism turn out to be correct. For in that case, the only a priori constraints that rational opinion must respect will concern the purely categorical, nonmodal structure of the world. And so we may proceed as follows: describe a hypothetical situation; specify the categorical information about that situation available to some imagined perfectly rational agent; ask what, on the basis of that information and the purely categorical constraints on rational credence, she ought to believe; check to see whether this result conforms to our credence-chance principle. When we do so, I think we discover two reasons for doubting that, in a Humean setting, $(\star)$ is the correct such principle.

To bring the first reason into view, consider a T-junction world in which each particle possesses a value for some intrinsic physical parameter, which we'll call "mass."[26] Let us suppose, as usual, that there are a gazillion instances of particles entering T-junctions. What's more, the exit statistics display, to a very close approximation, a very simple dependency on mass; as a consequence, the best system assigns a chance of exiting left equal to e^{-km} (for some constant k). Suppose our rational agent knows all this; that is, she knows that the exit statistics are such as to make an ur-chance function yielding this assignment "best." Next, she knows that a particle with mass 1 is about to enter a T-junction; so the best system assigns it a chance (say) of 0.1 of exiting left. And finally, she knows that this situation is a bit of a statistical outlier, in that in no other case does a particle with mass 1 ever enter a T-junction. What ought her credence be that this particle will exit left?

One answer is given by a principle of indifference: she knows that there are just two possible outcomes, so she should assign credence 0.5 to each. Another answer is given by (★): she should assign credence 0.1. And of course there might be other answers. The question before us is why, given Humeanism, the answer provided by (★) should trump all others, and in particular the answer given by an indifference principle.

To appreciate the question's difficulty, imagine our agent's information coming in a certain order. First she learns that whenever a particle enters a T-junction, it exits either left or right. Then she learns that in all intrinsic respects, this particular instance of a particle entering a T-junction is the only one of its kind. So far, it is at least rationally permissible – and quite possibly rationally obligatory – for her to assign a credence of 0.5 to a left exit. Next, she acquires a very general kind of information about other T-junction interactions: she learns that for very many values m of mass, very many particles with mass m enter T-junctions, and very close to e^{-km} of them exit left. What remains unclear[27] is why, exactly, this new information should have any bearing on her credence concerning the instance she is considering – indeed, a bearing specific enough that she should change her credence in a left exit from 0.5 to 0.1. I will leave this issue unresolved, save to note that it can't be handled this way: the extra statistical evidence available to our agent provides her with excellent reason for judging that T-junction interactions are *governed by a probabilistic law*; that law assigns a chance of 0.1 to a left exit on this occasion; so she should conform her credence in a left exit to this chance. That won't do, in a Humean setting, since no "governing" is going on. There is no conceptual distance, in this setting, between a probabilistic law and the high level summary of statistical behavior we have already stipulated our agent to possess. So what we need to understand is how this high level summary can have the very specific evidential force required by (★) *without* mediation through an inference concerning "underlying" chances. That, I submit, remains a mystery.

There is a second, closely connected mystery, one which may be much harder to solve – at least, if we insist on trying to make Humeanism compatible with principle (★). The credit for spotting this mystery goes to Ittay Nissan-Rozen, who in a very important recent paper (2017) argues that a credence-chance principle suitable for Humeanism must make use of a notion of admissibility, and – more surprisingly still – there will be cases where purely historical information is not admissible. While I will set up the difficulty somewhat differently, I want to gratefully acknowledge Nissan-Rozen's quite powerful insight (and encourage you, the reader, to consult his paper for a careful, detailed development of it and exploration of its consequences).

• Imagine a rational agent, convinced of the truth of Humeanism, living in a T-junction world that is also replete with Mazes. One hypothesis open to her to consider is that the best system for this world, *written in the language of fundamental physics*, asserts that each particle has a chance of exactly 0.5 of exiting left from any T-junction it enters, and that these events are all independent of one another. More generally, she might entertain the hypothesis that the fundamental laws of her world are L, where these laws entail the foregoing claim about the chance behavior of particles in T-junctions.

Consider some time t. Let the proposition B state that, at t, a certain particle will enter a certain Maze. Let the proposition A state that this particle will eventually exit via that Maze's dot 5. Let E be our rational agent's total evidence, at time t. Let Cr be a reasonable initial credence function such that her time-t credence is just $Cr(- \mid E)$. And now let us ask: what is $Cr(A \mid E \,\&\, B \,\&\, L)$?

That should be an easy question to answer. L and B together entail that the *chance* that the particle exits at dot 5 is ½.[28] The example involves no threat of 'undermining'; so − even for a Humean − this conditional credence should likewise be ½.

But matters are not so simple. Suppose that our agent's evidence includes the results of a massive amount of investigation of the behavior of particles in Mazes. That evidence has revealed an extraordinarily stable statistical pattern: about 27.8% of particles exit from dot 5, with no discernible correlations between different 'trials.' Nothing about the conjunction (E&B) distinguishes *this* trial as in any way different from the ones about which our agent has so much evidence. So, since our agent is rational − in particular, perfectly able to *learn from experience* − it must be that $Cr(A \mid E \,\&\, B) \approx 0.278$. Why should $Cr(A \mid E \,\&\, B \,\&\, L)$ be any different?

The anti-Humean has an obvious, straightforward answer. Given L, the past behavior of particles in Mazes can only be a massive fluke: something *allowed* by the laws, but judged by them to be extraordinarily objectively improbable. Fluke that it is, this behavior ought not − *given the truth of L* − to constrain rational credence about the behavior of the particle described by B. In short, adding L to what is being conditionalized on *undermines* the evidential significance that E would otherwise have for A.

But our imagined rational agent − convinced, remember, of the truth of Humeanism − cannot avail herself of this answer. One way to see why not draws on our earlier discussion of imperialism. Suppose we grant that a candidate system may draw on distinct vocabularies, and thus may specify more than one ur-chance function. In the present case, one of these functions is picked out in the language of T-junctions; *that* function assigns a chance of 0.5 to the proposition A. But for all our agent knows, the best system for her world *also* specifies − this time, in the language of Mazes − a function that assigns a chance of 0.278 to A. Supposing it does, then while it is true that *one* of the chance-functions that captures statistical patterns in her world treats her evidence as "flukish," there is *another* such function that doesn't. So without additional premises, we can't just explain away the difference between $Cr(A \mid E \,\&\, B)$ and $Cr(A \mid E \,\&\, B \,\&\, L)$ by claiming that, conditional on L, her evidence E becomes irrelevant-because-flukish.

In fact, the underlying issue remains, even if we build imperialism into our Humean account of chance. Suppose, for example, that we decide on reflection that it's an analytic truth about "objective chance" that there is at most *one* ur-chance function per world; we might then insist that a best system may specify at most one ur-chance function, and must do so in a language suitable to capture the fundamental micro-physics of the given world. In the present case, then, this version of Humeanism will only recognize an objective chance of 0.5 for A. But even if this is the version of Humeanism our imagined rational agent endorses, that makes absolutely no difference to the underlying epistemological issue. Consider that in order to be able to learn from experience at all, her credence function must have various biases built into it – roughly, she must treat it as antecedently highly likely that she lives in a suitably "uniform" world. What we confront, in the present case, is a sharp question as to what "uniform" means – and whether it can mean what it *needs* to, in order to secure the result that $Cr(A \mid E \& B \& L) = 0.5$.

To bring this question into focus, let's begin with a very plausible assumption about the nature of some of the rational biases built into our agent's credence function. Let L^- be the *nonprobabilistic* part of L; that is, L^- says that it is a fundamental law that there are particles moving in such-and-such ways, where the "such-and-such" includes the information that every particle entering a T-junction exits either to the left or to the right, but L^- specifies no ur-chance function. Then we may assume that our agent considers it overwhelmingly likely, *given* L^-, that any two sufficiently large, reasonably naturally shaped regions of spacetime feature very similar statistics in the behavior of particles in T-junctions.[29] Now, what L *adds* to L^- is, in effect, the information that the simplest probabilistic summary of the global exit statistics assigns a uniform chance of 0.5 to each left-exit, and treats these events as independent. So our agent should, plausibly, consider it overwhelmingly likely, *given* L, that in any sufficiently large, reasonably naturally shaped region of spacetime, the fraction of left-exits will be close to 0.5, and, more to the point, that there will be no correlations between exits *that are simple to specify in the language of T-junctions* (i.e., in the language suitable for directly describing the world at the most fundamental microphysical level).

But there will – of *course* – be correlations galore, that are *complicated* to specify in this language. And we've already stipulated that there is no simple way to describe Mazes in this language. So it is perfectly open to add the further hypothesis M that there are just those incredibly-complicated-to-specify (in the "fundamental" language) correlations needed to make it the case that (i) the fraction of dot-5-exits from Mazes (over all of space and time) is 0.278; (ii) there are no simple-to-specify correlations in Maze behavior (simple to specify, that is, in the language of Mazes). Perfectly open, that is, in the following epistemic sense: our agent's rational bias toward believing that nature is uniform at the level of T-junction behavior is not in the least threatened, if she comes to believe (or give a very high credence to) – specifically, on the basis of her considerable store of evidence E – the hypothesis M. So what sort of irrationality could she be guilty of, if $Cr(A \mid E \& B \& L) = 0.278$?

That's not a rhetorical question. Indeed, the first point to emphasize is that the rational biases just described *can't* be the whole story about what grounds the rationality of our agent's credences; just observe that she can conform to these biases while, for example, considering it certain that the first T-junction exit that happens every Monday is a left-exit. So we already knew that there must be further constraints on her credences, in virtue of which they count as rational. Perhaps it is one or more of these further constraints that guarantees that, on pain of irrationality, $Cr(A \mid E \& B \& L)$ must equal 0.5.

But there is a real puzzle about what these further constraints could be – what, in detail, a Humean epistemology could look like that will deliver the result that $Cr(A \mid E \& B \& L) = 0.5$. I'll consider and reject two options, and then leave it open where to look for more successful alternatives. The first option combines imperialism with a bald insistence that principle (⋆) just is one of the nonderivative, nondefeasible categorical constraints on rational credence. (You *need* to combine (⋆) with imperialism, else the principle becomes either inconsistent or too ambiguous to deliver the wanted result.) But we've already seen that imperialism is in some tension with Humeanism. And its combination with principle (⋆) raises an additional challenge: for we'll need to find some principled basis for settling which candidate ur-chance function is the "real" one, while simultaneously selecting the ur-chance function best suited to constrain credence in the way (⋆) requires. Finally, to simply invoke (⋆) as one of the fundamental categorical constraints on rational credence built into a Humean epistemology is to ignore the *explanatory* demand raised by the scenario we've been considering. It's clearly rational for our agent to have $Cr(A \mid E \& B) \approx 0.278$. (Stronger: it's *irrational* for her to have a substantially *different* credence.) Why is it, exactly, that conditionalizing on L requires her to change this credence to 0.5? A good answer should make it clear why a certain claim about statistical patterns visible at the level of T-junction behavior should render *irrelevant* her wealth of information about statistical patters visible at the level of Maze behavior. Baldly insisting on (⋆) as a nonderivative constraint on her credence simply punts on that demand.

A second, more interesting option invokes a kind of symmetry constraint on her credence (one similar to exchangeability). Here is a natural proposal about the form of such a constraint: Suppose worlds $w1$ and $w2$ both have L^- as the nonprobabilistic part of their best systems. And suppose they agree on their numbers of left-exits and right-exits from T-junctions. Then our agent's initial credence must be such that $Cr(w1) = Cr(w2)$. Given such a principle, we can plausibly argue not just that $Cr(A \mid E \& B \& L) = 0.5$ but, more strongly, that $Cr(A \mid E \& B \& L) = 0.5$.[30]

Unfortunately, this turns out to be too much of a good thing. For the derivation in fact shows that for *any* L^* that extends L^- by specifying some ur-chance function, correlations between T-junction behavior that L^* "sees" will make no difference to the agent's credence. Consider, for example, a world in which particles oscillate statistically in their T-junction behavior: it's true about 90% of the time that after a particle left-exits, its very next exit is a right-exit; likewise, it's true about 90%

of the time that after a particle right-exits, its very next exit is a left-exit. The best system for such a world will pick up on this pattern, and build it into the specification of the ur-chance function, by assigning to each particle a chance of 0.9 of "flipping" from one exit to the next. Let $L^\star$ be this best system. Let X be the proposition that a certain particle has just executed a left-exit. Let *left-right* be the proposition that its next two exits, in order, will be left, then right; left *right-left* be the proposition that its next two exits, in order, will be right, then left. Principle $(\star)$ yields the result that *Cr* (*left-right* | X & $L^\star$) = (0.1)•(0.9), while *Cr* (*right-left* | X & $L^\star$) = (0.9)•(0.9). But our symmetry principle yields the result that these two conditional credences must be equal. So far from providing a basis for deriving $(\star)$, our symmetry principle in fact conflicts with it.

At this point, it may be that the best hope for a plausible Humean epistemology of chance lies not in trying to vindicate principle $(\star)$, but in modifying it — specifically so as to allow that $Cr(A \mid E \& B \& L) = Cr(A \mid E \& B) \approx 0.278$. That is Nissan-Rozen's important suggestion: we should return to Lewis' original idea that an agent's beliefs about chances ought to guide her credences, *provided* she possesses (or is conditionalizing upon) no inadmissible information; and we should recognize that in the present case, the agent's knowledge of past Maze behavior counts as inadmissible information.

Now, principle $(\star)$ was already written in a way that allowed it to accommodate *ordinary* sorts of inadmissible information. Suppose, for example, that A is the proposition that a certain particle that entered a T-junction at time t exited left from it. Suppose it's a bit later, and our agent's total evidence E includes the proposition that 58% of this particle's exits from time t on have been left-exits (and includes nothing else relevant). Suppose she knows that the chance of each left-exit is exactly 0.5. Even so, her credence in A should surely be 0.58, and not 0.5. So it can seem, naively, that her beliefs about chances don't constrain her credence in this case, thanks to her possession of inadmissible information.

Not so. Look again at the statement of $(\star)$:

Let A and E be any two propositions. Let Pr be some probability function, and let X be the proposition true at a world w iff w's ur-chance function is Pr. Then

$$Cr(A \mid E \& X) = \Pr(A \mid E \& X) \tag{$\star$}$$

Applied to this case, we simply get the result that her credence in A should be set by (what she knows to be) the *conditional* chance of A, given E. And this conditional chance will equal 0.58. The key idea is quite simple: the relevance of the agent's evidence is, at it were, simply reflected in the structure of the conditional objective chances.

What Nissan-Rozen points out is that matters are fundamentally different, in the case we have been considering. Crucially, if Pr is the ur-chance function picked out at the level of T-junction behavior — that is, the function that assigns to each left-exit a constant and independent chance of 0.5 — then, since A concerns an outcome that is wholly distinct from the stretch of history that the evidence E is about, $\Pr(A \mid E \& X) = \Pr(A \mid X) = P(A) = 0.5$. The *agent* sees (quite properly) that her evidence bears on A; but the *chances* do not.

That means that (⋆) itself must be qualified by an admissibility requirement (bracketing the option of outright rejecting it): it describes a constraint that our agent's credences must respect, *provided* that the proposition E contains no information that is inadmissible with respect to the given proposition A and chances Pr. The intuitive idea, as we've just seen, is that E is inadmissible just in case Pr fails to recognize its evidential bearing on A. Obviously, the crucial piece of unfinished business here is to replace this intuitive idea with something more precise, and more illuminating.

9.7 Coming Clean

We've seen two respects in which a Humean approach to chance departs dramatically from an anti-Humean approach – and *not* just in its metaphysical foundations. There are more such respects: Humeanism will, for example, imply very different conclusions about the relationship between chance and time, laws, explanation, independence principles, and more.[31] So while I do not think we are as yet in a position to determine which of Humeanism and anti-Humeanism is the *right* approach, I *do* think that we're only going to make progress on that question if, as it were, Humeanism *comes clean*. We already knew that Humeanism is squarely at odds with basic modal intuitions about chance that most of us share: just consider the simple thought that exactly the same statistical pattern can be generated by different underlying probabilities. But what I've tried to indicate in this essay is that Humeanism about chance is much more starkly revisionary than *that*. And unless and until we understand the full contours of that revisionism, we won't be able to intelligently assess on which side of the great divide the truth about chance rests.

Notes

1 When the set S is non-denumerable, we do the usual thing: define urch$_w$ over an algebra of subsets of S. Also, we will see below some reason to think that worlds may have *multiple* ur-chance functions associated with them.
2 With, of course, the usual axioms in place: these probabilities are non-negative, and sum to 1.
3 In a relativistic setting: iff there is some Cauchy surface C in w and C⋆ in v such that the portion of w to the past of C is exactly qualitatively the same as the portion of v to the past of C⋆.
4 In a relativistic setting, we need to work a bit harder to get this relation to be an equivalence relation. One approach: assume that there is some way of partitioning each world in S into a set of non-overlapping but exhaustive Cauchy surfaces. In the definition of "starts out the same" given in the last footnote, restrict the choice of Cauchy surfaces to *these* ones.

5 To properly capture the idea that facts about physical modalities *reduce to* non-modal facts, we might want a stronger notion than supervenience; for example, perhaps the right account is that modal facts are *grounded in* non-modal facts. See, e.g., Rosen (2010).

6 For example, perhaps part of *what it is* for something to have mass is for it to attract other things with mass (see e.g. Ellis, 2002).

7 For example, perhaps laws consist in higher-order relations of necessitation between universals (see e.g. Armstrong, 1983).

8 Note one difference between the two examples. In explaining Kepler's second law, we pick out a range of facts some of which appear to have a *special metaphysical status*: they are laws of nature. (And yes, we should remember that for many philosophers, these appearances will be misleading.) By contrast, the fact that Suzy threw a rock at the window is hardly a metaphysically special sort of fact; what matters, rather, is that it bears (apparently!) a metaphysically special relationship (viz., *being a cause of*) to the event to be explained. In short, on a "special sort of information" view, explanatory information might itself be special, or might instead be ordinary information picked out because of its special relationship to the explanandum.

9 Many – most, really – contemporary accounts of scientific explanation fall under this broad approach. See for example Woodward (2005) and Strevens (2009).

10 For an excellent overview, see Ghirardi (2016).

11 And yes, "right" and "left" are shorthand for directions that are physically distinguished somehow by the structure of the T-junction.

12 Though see Armstrong (1983) for a more complicated anti-Humean account.

13 As in Hempel (1965b); note that this was early enough that the label "Humean" had not yet taken hold.

14 See for example Lewis (1994), Beebee (2000), Loewer (2004).

15 Note that for "simplicity" to be a useful criterion, we must impose some restriction on the *language* in which these sentences are expressed; else we can say anything we want in a perfectly simple manner, just by gerrymandering a language with suitable predicates. (See Lewis, 1983.) We'll return to this point below.

16 I.e., it conforms to the Kolmogorov axioms.

17 Note that I'm being very slightly misleading here, as Lewis did not distinguish sentences from measure in the way I have. But this is only a difference in presentation, not in substance.

18 Note, though, that these will be *conditional* probabilities: e.g., the probability that a certain particle left-exits a T-junction just after time t, given that it entered the junction at t, is 0.618.

19 Proof: First, it is enough to handle the case where f_1 and f_2 are both rational (with $f_1 \in (0, 1)$ and $f_2 \in [0, 1]$). When a particle enters a Maze, it will execute a certain number of "left loops" (left at the central T-junction, then left again) and a certain number of "right loops" (right at the central T-junction, then right again), in some order, before finally exiting the maze by going left at the central T-junction followed by right at the next junction, or right at the central junction followed by left at the next junction. Let A be the total number of left loops, B the total number of right loops. Let L be the total number of left exits, R the total number of right exits. Then $L = 2A + 1$; $R = 2B + 1$. We may assume that f_1 can be expressed as n/(n+m), where n is odd, and n and m share no common factors. (For if n is even, then we could just consider the frequency of *right*-exits instead.) There are now two cases. First case: m is odd as well. Then, since A and B can be freely chosen, we simply stipulate that, every

time a particle enters a Maze, $A = (n-1)/2$ and $B = (m-1)/2$. So in each such case, $L = n$ and $R = m$; it follows that the frequency of left-exits is $n/(n+m)$. Second case: m is even. This time, we stipulate that the total number of Maze entries is *even*, and that in exactly half of these cases $B = m/2$, while in the other half $B = (m-2)/2$; in *all* cases, $A = (n-1)/2$. Then for any pair of Maze entries (one of each type), $L = 2n$ and $R = m+1+m-1 = 2m$; so again, the frequency of left-exits is $n/(n+m)$. Finally, observe that the frequency of dot-5-exits depends *only* on the order of the last two T-junction exits in the Mazes; so we can set it to any rational value we want, provided only that there are enough Maze entries.

20 Lewis (1983) suggested a quite severe restriction: there is really just one appropriate language, namely that language whose primitive vocabulary expresses the most perfectly natural properties and relations instantiated in the given world. For a much more permissive approach, see Callender & Cohen (2009).

21 Note that we have not said enough about what, precisely, distinguishes Mazes to settle which option obtains. All we've said is that every Maze contains within it a certain network of linked T-junctions and channels; but for all that, there may be much more, micro-physically speaking, to being a Maze.

22 With the caveat that there will likely be events to which the second ur-chance function assigns no value.

23 To begin the derivation, we'll assume that the t-chance of some proposition P is given by $\mathrm{urch}(P \mid S_t)$, where S_t is the proposition correctly and completely describing the state of the world at t. Now consider two partitions on the space of possible worlds: $\{X_i\}$ and $\{S_j\}$, where each S_j completely describes a possible time t state of the world, and X_i is true at a world iff that world's ur-chance function is Pr_i. (Purely for convenience, we'll assume both partitions are countable.) So

$$Cr(A \mid E \ \& \ ch_t(A) = x) = \Sigma_{ij} \ Cr(A \mid E \ \& \ ch_t(A) = x \ \& \ X_i \ \& \ S_j) \cdot Cr(X_i \ \& \ S_j \mid E \ \& \ ch_t(A) = x),$$

where the sum is taken over all combinations $(X_i \ \& \ S_j)$ such that

$$Cr(E \ \& \ ch_t(A) = x \ \& \ X_i \ \& \ S_j) \neq 0.$$

But any such combination must in fact entail that $ch_t(A) = x$, so

$$Cr(A \mid E \ \& \ ch_t(A) = x) = \Sigma_{ij} \ Cr(A \mid E \ \& \ X_i \ \& \ S_j) \cdot Cr(X_i \ \& \ S_j \mid E \ \& \ ch_t(A) = x) = \Sigma_{ij} \ Pr_i(A \mid E \ \& \ X_i \ \& \ S_j) \cdot Cr(X_i \ \& \ S_j \mid E \ \& \ ch_t(A) = x).$$

If we add the background assumptions that ur-chances always treat themselves as certain (i.e., there is no "undermining"), then

$$Pr_i(X_i) = 1,$$

hence

$$Pr_i(A \mid E \ \& \ X_i \ \& \ S_j) = Pr_i(A \mid E \ \& \ S_j).$$

And if we add a temporal screening off condition, then, since E is entirely about history prior to t,

$$Pr_i(A \mid E \ \& \ S_j) = Pr_i(A \mid S_j) = x.$$

Therefore,

$$Cr(A \mid E \,\&\, ch_t(A) = x) = \Sigma_{ij}\, x \star Cr(X_i \,\&\, S_j \mid E \,\&\, ch_t(A) = x) = x.$$

24 As to why it is likely a mistake – at least, by anti-Humean lights – here is one reason: We know, I think, that if anti-Humeanism is correct, then objective chances are perfectly precise; but there is no way to derive this conclusion from a credence-chance principle. For other reasons, see Arntzenius and Hall (2003).

25 Nor should we hastily assume – as Lewis seemed to – that it cannot provide such a justification. See the discussion of what I called "primitivist hypothetical frequentism" in my 2004, for an attempted proof of concept.

26 What follows recapitulates an argument I gave in my 2004.

27 *Genuinely* unclear; this isn't that annoying, cheap rhetorical use of the term.

28 And this, just from the fact that, according to L, the behavior of any particle at any T-junction is probabilistically independent of the behavior of any other particles at any other T-junctions.

29 The patches need to be large enough to include sufficiently many instances of particles entering T-junctions. And they should be reasonably naturally shaped, since our agent knows for certain that, unless all or almost all particles exit in just one direction, there will be many scattered, oddly shaped patches that each feature many exits but that differ dramatically in their statistics.

30 Divide the worlds in which (E & B & L⁻) holds into those in which A is true, and those in which A is false. Then we need a modest assumption: there is a one-one mapping between these two sets of worlds such that, when w1 is a world in the first set, its image under the mapping w2 features exactly the same numbers of left- and right-exits. (The idea is that we get from w1 to w2 by swapping the last two T-junction exits inside the given Maze, and holding all other T-junction exits – whether inside or outside this Maze – fixed.) Then given our symmetry constraint on the agent's credence, it will follow that Cr(A & E & B & L⁻) = Cr(not-A & E & B & L⁻), hence Cr(A | E & B & L⁻) = 0.5.

31 See the writer's cut version of this paper for details.

References

Albert, David Z. (2000). *Time and chance*. Cambridge MA: Harvard University Press.

Albert, David Z. (2015a). *After physics*. Cambridge MA: Harvard University Press.

Albert, David Z. (2015b). Physics and chance. In *After physics* (pp. 1–30). Cambridge, MA: Harvard University Press.

Armstrong, D. M. (1983). *What is a law of nature*. New York: Cambridge University Press.

Arntzenius, Frank, & Hall, Ned. (2003). On what we know about chance. *British Journal for the Philosophy of Science, 54*, 171–179.

Beebee, Helen. (2000). The non-governing conception of laws of nature. *Philosophy and Phenomenological Research, 61*, 571–593.

Callender, Craig, & Cohen, Jonathan. (2009). A better best system account of lawhood. *Philosophical Studies, 145*, 1–34.

Carnap, Rudolf. (1945). The two concepts of probability. *Philosophy and Phenomenological Research, 5*, 513–532.

Ellis, Brian. (2002). *The philosophy of nature: a guide to the new ssentialism*. Oxford: Routledge.

Ghirardi, Giancarlo. (2016). Collapse theories. In Edward N. Zalta (Ed.), *The Stanford Encyclopedia of Philosophy (Spring 2016 Edition)*. URL = <http://plato.stanford.edu/archives/spr2016/entries/qm-collapse/>

Hale, Bob, & Hoffmann, Aviv (Eds.). (2010). *Modality: metaphysics, logic, and epistemology*. Oxford: Oxford University Press.

Hall, Ned. (2010). *Humean reductionism about laws of nature*. Unpublished manuscript.

Hall, Ned. (2004). Two mistakes about credence and chance. *Australasian Journal of Philosophy*, *82*(1), 93–111.

Hempel, Carl. (1965a). *Aspects of scientific explanation and other essays in the philosophy of science*. New York: Free Press.

Hempel, Carl. (1965b). Aspects of scientific explanation. In *Aspects of scientific explanation and other essays in the philosophy of science* (pp. 331–496). New York: Free Press.

Hicks, Michael. (2014). *The epistemic role account of lawhood*. Unpublished manuscript.

Joyce, James. (2010). A defence of imprecise credences in inference and decision making. *Philosophical Perspectives*, *24*, 281–323.

Kitcher, Philip. (1989). Explanatory unification and the causal structure of the world. In P. Kitcher & W. Salmon (Eds.), University of Minnesota Press (Trans.), *Scientific explanation* (pp. 410–505). Minneapolis.

Lewis, David. (1986). *Philosophical papers, Volume II*. Oxford: Oxford University Press.

Lewis, David. (1994). Humean supervenience debugged. *Mind*, *103*(412), 473–490.

Lewis, David. (1983). New work for a theory of universals. *Australasian Journal of Philosophy*, *61*, 343–377.

Lewis, David. (1980). A Subjectivist's Guide to Objective Chance. In Lewis (1986, pp. 83–113).

Loewer, Barry. (2004). David Lewis's Humean theory of objective chance. *Philosophy of Science*, *71*(5), 1115–1125.

Maudlin, Tim. (2011). *Quantum non-locality and relativity* (3rd ed.). Oxford: Blackwell.

Maudlin, Tim. (2007). What could be objective about probabilities? *Studies in History and Philosophy of Science*, *38*(2), 275–291.

Nissan-Rozen, Ittay. (2017). On the inadmissibility of some historical information. *Philosophy and Phenomenological Research*, *97*(2), 479–493.

Rosen, Gideon. (2010). Metaphysical dependence: grounding and reduction. In Bob Hale & Aviv Hoffmann (Eds.), *Modality: metaphysics, logic, and epistemology* (pp. 109–136). Oxford: Oxford University Press.

Strevens, Michael. (2009). *Depth: an account of scientific explanation*. Cambridge, MA: Harvard University Press.

Woodward, James. (2005). *Making things happen: a theory of causal explanation*. Oxford: Oxford University Press.

Xia, Zhihong. (1992). The existence of noncollision singularities in Newtonian systems. *Annals of Mathematics*, *135*, 411–468.

10 On Chance (or, Why I am Only a Half-Humean)

J. T. Ismael

Before the development of quantum mechanics, most of the philosophical discussion of probability focused on statistical probabilities.[1] Philosophers of science have a particular interest in statistical probabilities because they play an important role in the testing and confirmation of theories, and they played a central role in the statistical mechanics of Boltzmann and Gibbs developed in the eighteenth century. Since the introduction of quantum mechanics, however, much of the philosophical attention has become focused on the interpretation of chances. These are the probabilities assigned to particular events (the detection of a photon at a certain location on a photographic plate, or the registration of the result of a spin experiment on a particular electron) by applications of the Born Rule. The appearance of chances in quantum mechanics marked the first time that probabilities made an explicit appearance in a *fundamental* theory. They raise new kinds of ontological questions. Unlike statistical probabilities (which pertain to classes of events), chances are single-case probabilities. And unlike credences (which represent the epistemic states of believers), chances purport to represent features of the physical world.

Hall's chapter (this volume) introduces the main divide in the philosophical discussion of chances and shows how the difference in orientation between Humean and anti-Humean views shapes the detailed development of those views. In this chapter, I defend a half-Humean view that retains the ontological thesis that motivates Humeanism, but denies that the Humean account does (or should) provide a content-preserving reduction of statements about chance to statements about nonchancy facts. The first part of the chapter is expository. In Sections 10.1 and 10.2, I sketch the history of Humean accounts of chance. In Section 10.3, I introduce the form of the contemporary Humean account. In Section 10.4, I introduce the two arguments against Humeanism that (in my assessment) cut closest to its philosophical heart. The second part of the chapter makes a positive contribution to the development of the Humean account. In Section 10.5, I situate chances in the matrix of probabilistic notions. In Section 10.6, I say why I think Humeans should not try to reduce chances to nonchancy facts, and in Section 10.7, I introduce the *half*-Humean view that I favor. This is an entry into a lively and ongoing discussion.[2] No attempt is made to provide a comprehensive survey of arguments for and against Humeanism. The reader is encouraged to look into some of the discussions indicated in the endnotes.

10.1 History

Questions about the nature of chance were part of the general ferment surrounding the interpretation of quantum mechanics in the foundations of physics for most of the twentieth century, and David Lewis propelled them to the forefront of metaphysician's attention with a 1980 paper (Lewis, 1980). One of the central questions in the metaphysics of science has always concerned the status of modal notions: e.g., laws, causes, dispositions, and capacities. At the time that Lewis was writing, the debate about these notions had settled into two broad classes of view: so-called Humean and anti-Humean views. The Humean holds that the world consists of what happens: just one thing and then another, arranged in a four-dimensional manifold of events, the totality of local matters of particular fact. According to the Humean, laws and chances are patterns in the manifold of events. For the anti-Humean, they *govern* and *explain* those patterns.[3] In Lewis' eyes, the program of Humean metaphysics hinged on the possibility of providing a Humean reduction of chances. He had developed a powerful framework for articulating the difference between the Humean and anti-Humean position and provided successful Humean reductions (by his lights and the lights of many of his followers) of laws and causation, but he despaired of providing a Humean reduction of chances. The result of this was that questions about the nature of chances aligned with perhaps the central dispute in the metaphysics of science, and it became a lightning rod for debate.

10.2 The Big Bad Bug

Here is how Lewis framed the issue. He assumed nothing about what the chances are quantitatively. In his mind, it was the purview of physics to tell us what values the chances take at different points in space and time, just as it was the purview of physics to tell us the values of the electromagnetic fields. But it was the purview of metaphysics to try to understand what sorts of things chance *are*. Are chances objective features of the physical world? Do they supervene on nonmodal facts, or are they fundamental features in their own right. How do they fit into the catalogue of Being? He introduced a principle that he thought expressed a connection to belief which provided everything that we know pre-theoretically about chance. He called it the Principal Principle (PP). The task for the Humean, as he saw it, was to find something that supervenes on the collection of local matters of fact, which could play the role of chance guiding belief expressed by PP. The problem was that he thought it couldn't be done. The reason was laid out in a quite lovely argument in his 1980 paper. In the time since its publication, a massive literature has built up around the various ways of addressing Lewis' worry. A number of solutions have been proposed, including one that Lewis himself accepted (Lewis, 1994).[4] And, since the issue touches on so many debates in the metaphysics of science, those discussions have been quite generally sharpened and deepened.

10.3 The Canonical Form of the Humean View: the Best Systems Analysis

The philosophical motivation for the Humean view, as noted by Hall, is a metaphysical vision of the world, combined with a view about the epistemic role that beliefs about chance play. David Albert, one of the most influential contemporary defenders of Humeanism, puts it thus:

> [On the Humean view] the world, considered as a whole, is merely, purely, there. It isn't the sort of thing that is susceptible of being explained or accounted for or traced back to something else. There isn't anything that it obeys. There is nothing to talk about over and above the totality of concrete particular facts. And science is in the business of producing the most compact and informative possible summary of that totality.
>
> (Albert, 2015, pp. 23–24)

The canonical form of the Humean view is given by (what has come to be known as) the Best Systems Analysis. We are told that beliefs about laws and chances come in packages ("best systems") that are chosen on the basis of simplicity, strength, and best overall fit with the Humean mosaic. Beliefs about laws and chances are products of the systematization of information about the Humean mosaic. The patterns in the Humean mosaic that provide the basis for choice between different systems are then presented as truth-makers for chance and law assertions. On this view statements about chance are compact summaries of information about distributed patterns in the manifold of events. The function of this kind of compact summary is to provide limited creatures information that will guide action and belief in a world too complex to be fully comprehended in a description we could grasp.

There is a lot of room under the Humean umbrella for different accounts of what makes a system a good one, and whether there is a single system for all of science, or many systems, one for each special science.[5] The Humean project can be thought of as a schema to be completed by providing an explicit account of how systematizations are chosen and individuated.[6]

There are open issues in the development of the program. The most important of these is that nobody has given an adequate, explicit account of what simplicity and strength are. Lewis himself tied the measure of simplicity and strength to his account of natural predicates, a part of his metaphysics that most contemporary Humeans among philosophers of science would rather do without.[7] It remains an outstanding task for the Humean account to fill this hole.[8] With that said, the Humean view has many proponents in the philosophy of science. It does a good job of capturing the function of scientific theories: viz., to systematize information about the Humean mosaic in a compact form for use by limited agents. It achieves a good match between the *function* of a scientific theory and the standards by which scientific theories are judged. Simplicity, strength and fit make good sense as standards by which theories are judged against one another if the goal is to systematize information about the Humean mosaic in a compact form. And it doesn't come with metaphysical commitments that seem at odds with an empiricist orientation. In sum, the Humean offers an account of how (chance+law) packages are formed that is meant to reproduce the epistemology of science, and she asserts that there is nothing more to being a law than being a theorem of the best system, and nothing more to being the correct distribution of chances than being the distribution entailed by the best system. The result is a metaphysically conservative view that holds that the fundamental modal postulates of a theory – the laws and the chance distribution – are nothing more than compact statements that encode information about the actual pattern of events. There are some outstanding questions that need to be answered in the development of the Humean account, but it remains a live program with many adherents.

10.4 Two Arguments against Humeanism

There are two quite powerful objections against Humeanism. The first one alleges that the Humean account robs the chances of explanatory power. It holds that chances are a substantial ontological posit needed in the explanation of why there are stabilized relative frequencies. The argument here parallels the argument for believing in anti-Humean laws. The claim is that without laws as substantial ontological posits, there is no explanation for all of the regularity in the world. What keeps the planets in orbit, and airplanes from falling out of the sky? If not the laws, then what? What keeps casinos in business and insurance companies making money? If not the chances, then what? Laws and chances, according to the anti-Humean, govern events and explain regularities. She will point to instances of scientific explanation that invoke laws and chances in an explanatory role.[9] She will bolster her case by urging that we have no better guide to metaphysics than physics. If laws and chances are primitive elements in our physical theories, we should treat them as a primitive part of our ontology.[10] She sees no reason for seeking reductions. She goes on to point out that in the internal logic of a theory laws and chances are invoked in explanations of phenomena (Emery, 2015). In sum, the anti-Humean holds that Humean chances can't play the explanatory role that chances play. She further questions the Humean methodology, asking what warrant there could be for a *meta*-physical viewpoint for reducing quantities that are treated as primitive within our best physical theories. Where the Humean says that laws and chances describe patterns in the manifold of fact, the anti-Humean says that laws and chances are independent existences that *govern* and *explain* the pattern of fact. According to her, theorizing is all aimed at identifying the explanatory substructure behind the facts.

The second anti-Humean argument contends that the BSA is the development of an old tradition that tries to reduce probabilities to frequencies, and holds that the BSA suffers from the same problem that the simpler accounts in that tradition suffer from. The frequency theorist says that probabilities are to be identified with frequencies that satisfy some criterion C (let's call these the C-frequencies), and the anti-reductionist points out that the law of large numbers assigns a nonzero probability to the possibility that the C-frequencies diverge from the probabilities. That seems to be true no matter what one fills in for C, and it presents an obstacle to identifying beliefs about chances with beliefs about frequencies of any kind.[11] The holistic reduction of (chance + law) *packages* to patterns in the manifold of actual events makes it hard to apply the law of large numbers directly, but it still seems to suffer from a version of this problem. If we look at the modal implications of accepting a system B as the Best System, we will find ourselves committed to the possibility of worlds in which the laws and chances aren't given by the best system *at that world*. So, for example, consider a simple world w, which consists of a sequence of a very large sequence of flips roughly half of which come up heads and half of which come up tails, in a pattern that doesn't admit of any compression. And suppose that the best systematization (B) of the results of the flips at w assigns 50% chance to heads on any given toss. Now, B has a model in which every toss comes up heads. The best systematization of the facts of that world would assign a 100% chance to heads. From the point of view of B, that is a lucky accident, but it is one whose possibility is explicitly recognized by B. It is hard not to speculate that any Best System (i.e., any package of laws and chances that systematizes the facts at a world of complexity close to ours) is going to have models in which nothing of much interest happens and which permits of a simpler systematization. So just as the very logic of probabilistic belief, as expressed by the law of large numbers, explicitly recognizes an ineliminable modal gap between probabilities and frequencies (of any kind that can be explicitly characterized by a criterion C), so it seems that the logic of Best Systems makes room for the possibility of worlds in which the laws and chances are not given by the best system (at that world).

The first of these arguments is an expression of what Hall calls the difference in orientation between Humean and anti-Humean views. It is question begging as an argument against Humeanism, because the Humean simply rejects the explanatory demand. The second argument is not easily dismissed. It is one of a family of arguments that purport to show on nonpartisan grounds that the Humean account of chance fails to provide a reduction of chances to nonchancy facts by showing that the truth conditions for statements about chance have a modal component that can't be paraphrased by any set of statements about purely categorical facts.[12] I think that the conclusion of these arguments is correct. In the remainder of the paper, I want to propose a direction in which the Humean view should be developed.[13] The suggestion will be that we should accept the modal commitments of accepting a Best System, but give an account of them that understands them purely in terms of their epistemic role.

10.5 The Matrix of Probabilistic Notions

To pave the way for that, there is another adjustment to be made to the Humean program. The strategy for Humeans and anti-Humeans alike has been, for the most part, to attack the question of what chances are directly. Since chances appear along-side laws in our fundamental theories, the presumption has been that they are the most fundamental form of objective probability. The task has been conceived as a matter of sorting out the connection between the chances and categorical facts directly, while imposing the connection to credence captured in PP (or one of its proposed variants) as a constraint. If we situate chances in the more complex matrix of probabilistic notions, however, we get a more nuanced understanding of how chances relate to the categorical facts, and that paves the way for the Half-Humean view I want to propose.[14] That is what I am going to do in this section. The discussion will be condensed, but details can be found in references.

A probability measure is a function, P[.], that maps events in the sample space, S, to real numbers such that:

P[A] ≥ 0 for any event A.

P[S] = 1.

$P[A_1 \cup A_2 \cup \ldots] = P[A_1]+P[A_2]+\ldots$ for any countable collection of mutually exclusive events $A_1, A_2\ldots$

Conditional probability of the event A given the occurrence of the event B is $P[A|B] = P[AB] / P[B]$ (AB: short for $A \cap B$).

We distinguish general probabilities from single-case probabilities. General probabilities pertain to classes and the basic form is conditional. Single-case probabilities (e.g., the probability that a particular carrier of the BRCA2 gene will develop cancer) pertain to individual events and the basic form is unconditional. General probabilities are related to single case probabilities by the principle that the single case probability of an x that is randomly selected from the population of y's is the general probability of x, given y. These definitions license an inference from a general probability to a single case probability where the selection procedure is random, and where nothing else is known that might affect the probability. So, for example, if the general probability that a person will develop cancer given that they carry the BRCA2 gene is 0.45, then the single case probability that a particular carrier of that gene will develop cancer is (ceteris paribus) 0.45.

Statistical probabilities are general probabilities. What makes them *probabilities* is that they obey the probability axioms. What makes them statistical is a connection to statistics. The connection to statistics comes from the law of large numbers (in its weak or strong version) which says that the relative frequency of x in a class of randomly selected y's approaches the general probability of x, given y, with increasing probability as the class gets larger.

In operational terms, statistical probabilities are elements in a theoretical matrix that mediate inference from observed statistics in local samples to others in the same class. Probabilistic notions are connected to actual observed frequencies by the body of operational procedures and norms by which we infer probabilities from collected statistics. The operational procedures are embodied in the canons of statistical inference.[15]

Assignment of statistical probabilities commits one to the expectation that the statistics for one sample will reflect those of others (provided that no selection was exercised either in the collection of the sample or in the target class, i.e., the one that we are forming expectations about).[16] The expectation grows as the size of the sample increases and is defeasible by the belief that the selection was biased or that the selection process was not random.

Statistical probabilities contrast with epistemic probabilities. Where statistical probabilities are interpreted by a connection to frequencies, epistemic probabilities are interpreted by a connection to belief. What makes them probabilities is (again) that they obey the probability axioms. What makes them epistemic is a connection to belief. There is a descriptive form of epistemic probabilities. These are the credences. They represent the epistemic states (or degrees of belief) of agents.

The Principal Principle (PP) was introduced by Lewis as an expression of the role that chance plays guiding belief. His informal statement of the connection to belief was that if you know that the chance of e is x, and you don't have a crystal ball, or any other form of supernatural information from the future, your credence in e should be x. Few would nowadays agree with Lewis that PP captures 'all we know about chance,' but almost everyone agrees that PP (or something quiet close to it) captures a connection between chance and credence that acts as a constraint on the interpretation of chance.[17] For that reason, the principle continues to play a central role in the philosophical discussion of the metaphysics of chance. PP tells us how chance is situated in this matrix. It tells us chances are a normative form of epistemic probability. And it tells us that the chances are adopted as credences in the absence of information from the future.[18]

Now the question is how do we connect this matrix to the categorical facts. We begin with the most direct point of contact; viz., the general statistical probabilities associated with stabilized relative frequencies across reference classes of the kind that casinos, lotteries, and insurance companies rely on.[19] These kinds of probabilities are defined for types rather than tokens, and the basic form is conditional. They exist in deterministic as well as indeterministic contexts. We have a probability for a/b when we have a relative frequency of a's among b's with the right kind of stability (i.e., where the frequency is roughly stable across not carefully chosen subselections from the b's).[20] We have good philosophical models of these kinds of probabilities in specific cases (the best known is the Diachonis model of the coin flip). The dynamical underpinnings, however, will vary from case to case, and the knowledge that there is a probability associated with an event in some reference class (i.e., a relative frequency with the right kind of microstructure) often precedes our explicit understanding of the dynamical underpinnings. Science was neck deep in these kinds of probabilities long before quantum mechanics came on the scene, and probabilistic thinking in everyday life relies on the existence of these kinds of emergent stabilized relative frequencies that let us form reasonable expectations where we have only partial knowledge.[21]

We can sketch a broadly Humean story about the emergence and use of probabilistic thinking. Humeans think that there are physical systems and their categorical properties. They think that there are laws that determine the physically allowable trajectories. So far, there is nothing 'probability-like.'

There is a large body of work shows us how to introduce **general conditional probabilities of the form pr(A/B)** where we have stabilized relative frequencies that support projection to new random sub-selections from the relevant class.[22] These general conditional probabilities don't have the right form to play the role guiding belief characterized in PP, precisely because they are *general* and *conditional*. PP requires something that is single-case and unconditional. For any given event – say the event that a particular coin toss comes up heads – there are indefinitely many general probabilities that apply to it. Choose any reference class to which the toss belongs and provided the class has the kind of microstructure that allows us to attach a probability, we will have the probability that coin tosses in that class come up heads, and they will not in general all be the same. This is, of course, the old reference class problem, which PP avoids only because it is solved in the step from general probabilities to chances. First we find which of those general probabilities fixes the chance, and then we use the chances, via PP, to fix credence.

Once we have the general probabilities we can look to identify the **single-case, unconditional probability** of an event derived from the general conditional probability that can play that role, and so here we just look at what PP says. PP says adopt *chances* as credences no matter what information you have from the past, provided you have no information from the future, so we look for probabilities that screen off all and only information from the past.[23]

There are two natural candidates:

(i) The first goes naturally with a Lewisian framework in which theories of chance take the form of history-to-chance conditionals, the chance of e at t =[24] the general probability of an event like e following a pre-t history (where both e and pre-t history are characterized in intrinsic, or qualitative, terms). There are different ways that we might think about this quantity: e.g., as an expression of the information that history contains about e, or as a measure of the propensity of pre-t histories to produce e.

(ii) The second fits more naturally with a physical setting in which we think of the intrinsic state of a system as determining the probabilities of events that fall in its future. In this case, we conditionalize not on all of history, but on the intrinsic state of the system in whose future e lies, so the chance of e for a system in a state S= the general probability of an e-type event for a system whose intrinsic state is S. As above, we can think of this quantity either as an expression of the information that S's intrinsic state contains about whether e, or as a measure of the propensity of a system in S state to produce e.

If we assume a Markov condition, i.e., if intrinsic character of a system screens off any information from history, (i) and (ii) are quantitatively the same. Either of them will screen off historical information, but not information from the future (or, in a relativistic setting, information that is not drawn from the absolute past), and so either of them will be suited to playing the role of chance in guiding credence.

There are interesting connections between (i) and (ii). There's some very nice work in computational mechanics, for example, that shows that (under quite weak assumptions) we can start with a set of observable quantities and divide histories into equivalence classes that generate the same conditional probabilities for the observables. We can then use those to construct a Markov chain of what are called 'causal states' that screen off information from the past and generate the same conditional probabilities.[25] What the causal states capture is all of the intrinsic structure in the system that is relevant to predicting the variables of interest. So there are lots of interesting things to say about the relationship between these two options, but in the special case in which states are causal states in the sense above, either of them will screen off all and only information from the past, and so either of them will be well-suited to play the role of chance in PP.

Finally, there are **credences**, which are subjective degrees of belief. These are descriptive of the epistemic states of believers.

What we have above are three logically distinct notions of probability: (i) the **general conditional probabilities**, which we can link to stabilized relative frequencies in reference classes, (ii) the **chances** (single-case unconditional probabilities that guide belief in accordance with PP), and (iii) the **credences**. We can say some things about relationships among them. The relationship between the general conditional probabilities and the intrinsic state of the system to which they are assigned is metaphysically contingent in the sense that it is mediated by laws that relate the intrinsic state of a system to its future behavior. We can think of the categorical properties as the bearers of chances, at least in a world in which the present chances pertaining to S supervene on the intrinsic state of S. The connection between general conditional probabilities and credences is likewise contingent, mediated in this case by facts about our epistemic situation. You shouldn't adopt chances as credences if you regularly get information from the future. They are good guides to belief only for creatures like us who don't have crystal balls, i.e., whose information about the future all comes by way of information about the present and past.[26]

In sum, we have a matrix of probabilistic notions, connected at one end to belief and at the other end to stabilized relative frequencies. Chance is linked to belief by PP, and general probabilities are linked to stabilized relative frequencies by the law of large numbers. The internal links among elements in this matrix are uncontroversial except for the connection between chances and general probabilities,[27] which is constrained by the alignment of other elements.

10.6 The Prospects for Reduction

When chance is situated in this matrix, the closest connection that we get between chances and categorical facts comes from the link between general probabilities and stabilized relative frequencies. This secures an evidential relationship between frequencies and probabilities. But it also blocks reduction because it leaves open the *possibility* that the actual frequencies may diverge as far as you like from the probabilities. As I've said, I think that on this point, the Humean should concede that her account does not provide a reduction. The connection between probabilities and statistics *interpret* the probabilities without *reducing* them. Empirical content flows into the matrix through the connection between statistical probabilities and stabilized relative frequencies.

To acknowledge that the connection isn't tight enough to permit reduction, however, is not to commit one to a substantial anti-Humean ontology for chances. The Humean need only note that there is a difference between making statements about patterns in the mosaic and making a guess or venturing an inductive hypothesis based on those patterns. She can hold that the chances are best guesses based on patterns in the Humean mosaic, but deny that they are to be identified with those patterns. The difference matters because if we are trying to capture the *content of beliefs* about chance, then there are good reasons for denying that they are the same as beliefs about patterns in the Humean mosaic. For one thing, they fail the modal test for sameness of content: we can hold the pattern of events fixed in our imagination and imagine that the chances vary. For another, they are not intersubstitutable in inference: beliefs about chances license modal beliefs, which beliefs about patterns in the Humean mosaic do not. There is an ineliminable logical gap between beliefs about the chances and beliefs about the pattern of categorical fact that is codified in the axioms of probability and also essential to the role of beliefs about chance in guiding expectation. That looseness of fit makes inferences from patterns in the Humean mosaic to chances unavoidably *ampliative*.

The Best Systems Analysis is best thought of not as an ontological reduction, but as a complicated inductive procedure that tries to extract patterns from what we know about the Humean mosaic and use them to guide credence about the future. This preserves the heart of the Humean idea that the modal structures are best thought of as epistemic guides. The idea is that a physical theory is a kind of inductive machine that guides prediction in the face of ignorance. One finds this idea in the writings of many Humeans. Carl Hoefer, for example, defending a Humean view with very strong affinities to the one here, echoes the idea that the chances are to be thought of as expressive of best guesses about unobserved cases based on stabilized relative frequencies.[28] He introduces as an example of a regularity,

> A page of a high school yearbook containing row after row of photos of 18 year olds, in alphabetical order — so that, in the large, there is a stable ratio of girl photos to boy photos on each page, say 25 girls to 23 boys). ... The regularity about boys and girls on pages is objectively there, and makes it reasonable to bet "girl" if offered a wager on the sex of a person whose photo will be chosen at random on a randomly selected page.
>
> (Hoefer, 2007, p. 18)

And of course he's right, the regularity is there in the world as part of the mosaic of categorical fact, but the probability has inductive content that goes beyond the regularity. It bets on its persistence, commits itself to the expectation of its persistence, and has modal implications that outrun any mere belief in how things are. It is certainly no logical contradiction to notice the regularity and not associate it with a probability, i.e., not to take on the additional commitments that come with an assignment of probability. One can easily imagine situations in which that would be the rational thing to do (so, for example, if you are sitting at a roulette table and you notice that the apparently random behavior is actually carefully managed by a lever under the table). So what is 'really there' in the world is the regularity. The inductive content is a hedged prediction *based on* the regularity that carries epistemic and practical commitments that we make explicit by looking at its role guiding expectation and betting behavior.[29]

If one goes back and looks at Lewis' papers, there are actually two separable threads in that paper: one that treats chances as recommended credences and one that identifies chances with patterns in the manifold of categorical fact.

My suggestion is that what the Humean view can be thought of is presenting the patterns in the Humean mosaic that provide the basis for inductive content built into the probabilistic belief. That inductive content doesn't make a full-on prediction about what will happen, but a hedged guess about how likely the alternatives are, given the system's intrinsic state and its history. The same goes for beliefs about laws, capacities, dispositions, and causes – i.e., all of the modal outputs of Best System style theorizing. These aren't first order beliefs about what is the case, but derivative quantities that encode inductive content based on patterns in the manifold of fact. Those quantities are *used* in epistemic and practical reasoning. Taking patterns in the mosaic and 'solving for the probabilities' is a way of preparing credences in advance for believers who find themselves with no direct source of information about the future but needing to make guesses. It is science doing what science does best: taking everyday informal inductive practices and making an art out of them. Beliefs about chance are hedged predictions. They don't assert, "this will happen" or even "this will happen so and so many times", but a generalized form of prediction that I've called elsewhere a 'partially prepared solution to a frequently encountered problem.' They extract projectible regularities from the pattern of fact and give us belief-forming and decision-making *policies* that have a general, pragmatic justification.

I propose that the right way to develop a Humean account is to hang onto the epistemic thread in Lewis' presentation as expressive of the content of statements about chance, and present the ontic part really as an account of the facts that ground the inductive guesses expressed by those statements. That accords quite well with the spirit of the Humean view.

10.7 The Half-Humean View

This kind of account departs from the more familiar Humeanism in two respects: (i) it doesn't treat chance as the most basic form of objective probability, and (ii) it denies that beliefs about chances (and laws) are simply disguised beliefs about patterns in the Humean mosaic. But it preserves the Humean ontology, remaining opposed to the reification of laws and chances as agents in the production of the phenomena. I would describe that view as giving a nonreductive analysis of modal *content* without inflating modal *ontology*. The half-Humean holds that chances are inductions on stabilized relative frequencies that go beyond a mere *description* of actual frequencies in any form, and that we can tell the story about the emergence and function of probabilistic thinking without invoking anything but stabilized relative frequencies, and hence without invoking anti-Humean truth makers for probabilistic belief.

But the anti-Humean will say that the Humean misses the point. She will say that we don't just need an account of how we form those beliefs; we need an account of what *warrants* them. We need something in the world to ground the inductive inferences that the Humean says are expressed in beliefs about chances. She will say that our lives depend quite literally on the continuation of lawful regularities in their strict or probabilistic form, and hold that unless laws and chances are agents in the production of behavior, we have no reason to expect regularities to continue. This is really the crux of the dispute between the Humean and anti-Humean. Our physical theories provide us with models with a good deal of epistemic superstructure that guides limited creatures through a complex world. That is why we build theories and that is how we use them. Both sides agree on that. The debate between the Humean and anti-Humean is about whether that epistemic superstructure is *nothing more* than epistemic superstructure, or whether it represents modal structure that is part of the intrinsic fabric of the physical world.

10.8 Conclusion

The divide between the Humean and anti-Humean accounts is deeply entrenched, with many able defenders on both sides. While my own allegiance is to the Humean view, I think that what divides the Humean and anti-Humean are quite deep issues about where demands for explanation bottom out, and what counts as an explanation. The Humean says that we can explain the formation and function of beliefs about chance without supposing the existence of anything but stabilized relative frequencies in the pattern of actual fact. The anti-Humean asks 'what supports the stabilized relative frequencies?' The Humean says 'all chains of explanation end somewhere,' or she says that if the anti-Humean thinks that simply postulating anti-Humean whatnots to keep the world running like clockwork really explains anything, she is wrong…. and (as Lewis might himself have said) so it goes.

Notes

1 The title of the chapter is a nod to Pearl (2001). An early version of this chapter was presented at a workshop in London organized by Mauricio Suarez. Richard Pettigrew, Luke Glynn, Seamus Bradley, Roman Frigg, and Nina Emery were present and I benefitted greatly from the discussion. I would especially like to thank Guido Bacciagaluppi for years of discussion. I've learned an enormous amount from him about chances and much else. Many thanks to Shamik Dasgupta for the invitation to contribute to this volume and for the most gracious possible editorship.
2 There may be other reasons for challenging the Humean ontology (Maudlin, 2007, pp. 78–103), but I put them aside here.
3 It was Lewis who coined the term 'Humean' in this capacity after Hume, the great denier of necessary connections between distinct events. Whether the labels is apt or not, it has stuck.
4 Much of this literature has focused on formulations of the Principal Principle. The strategy has been to reformulate PP in a way that gets around the problem that Lewis put his finger on. A thriving industry, which now spans the literature in metaphysics and formal epistemology, is devoted to these competing principles. See Hall (1994), Ismael (2008), (2015b), and Pettigrew (2015). The question of whether the Humean program could be reconciled with PP is widely regarded as having been settled in the affirmative, possibly with a little tweaking of the formal expression of PP.
5 See Cohen & Callendar (2009).

6   Here's Lewis's own classic statement of the method, in its application to laws: "Take all deductive systems whose theorems are true. Some are simpler better systematized than others. Some are stronger, more informative than others. These virtues compete: An uninformative system can be very simple; an unsystematized compendium of miscellaneous information can be very informative. The best system is the one that strikes as good a balance as truth will allow between simplicity and strength. How good a balance that is will depend on how kind nature is. A regularity is a law IFF it is a (contingent) theorem of the best system." (1994, p. 478)

The method was generalized by adding 'fit' to simplicity and strength to provide criteria for the choice between theories that include not only a set of laws but a theory of chance. So, says Lewis: "Consider deductive systems that pertain not only to what happens in history, but also to what the chances are of various outcomes in various situations - for instance the decay probabilities for atoms of various isotopes. Require these systems to be true in what they say about history.... Require also that these systems aren't in the business of guessing the outcomes of what, by their own lights, are chance events; they never say that A without also saying that A never had any chance of not coming about. (1994, p. 480) The idea is that the higher the chance a system assigns to the true history (or to segments of it given part of the history) the better its fit.

7   See Hall (2010) for a nice discussion of this aspect of Lewis's account.

8   See Loewer (2004), Callender and Cohen (2009). The difficulty isn't merely to provide measures of simplicity, strength and fit, but to provide measures that *capture criteria operative in scientific theory choice*. Lewis was clear that he wanted the Best Systems Analysis to reproduce the epistemology of science. Scientists routinely use words like 'simplicity,' 'strength,' and 'fit' to describe extra-empirical criteria for theory choice, but resolving the vague and qualitative character of those criteria into something precise and quantitative has proven elusive. One option for the Humean is to reject the demand for general univocal notions of simplicity, strength and fit that can be formally defined and slotted in as criteria for theory choice. The informal notions of simplicity, strength and fit provide some guidance about the standards for choosing models, but they are loose enough to allow the details to get filled in by looking at scientific practice and they allow for pragmatic trade-offs and some context- and problem-dependent choices of a kind one often finds in science. The pragmatic and metaphysically deflationary spirit of the Humean account allows for a little vagueness and variability in what the laws and chances are.

9   Emery, Nina (2015) provides a recent defense of both of these objections.

10   Maudlin (2007) has pushed this objection with some force.

11   Hajek (2012) contains a nice discussion of the difficulties of reduction.

12   Hall (this volume) mentions one. Maudlin (2007) presents one. Loewer (2004) responds that even though the simple world (S) is a model of our laws, our laws are not the laws of S. The laws of S are given by the best systematization of the facts at S. I don't think that the response is helpful. It strikes me that what these arguments are pointing to is that any (chance+law) package has a number of models and any Humean mosaic is a model of many different (chance+law) packages. This means that accepting a (chance+law) package comes with modal commitments that outrun any set of claims about the pattern of actual events. And that in its turn means that the content of a (chance+law) package is not exhausted by what it entails about the pattern of actual events. Whatever the nature of the inference from a Humean mosaic to a chance+law package is, I think that these arguments show that it falls short of reduction.

13   See Ismael (2015a).

14 There is another reason for wanting to do this; confusions generated by the way that the word 'chance' is used. In the tradition stemming from Lewis that conceives of a theory of chance as a collection of history-to-chance conditionals, and the chances as something derived from such a theory in a manner that is tailored to serve as a guide for belief. The philosophical tradition has mostly followed Lewis in using 'chance' to refer to the single-case probabilities. The foundational literature in physics uses 'chance' more loosely, sometimes in specific ways disambiguated by context, but often generically for any form of objective probability. And that usage has seeped into the philosophical literature in a way that produces confusion (e.g. Hoefer 1997, 2007; Albert, 2000).

15 There is no compact simple rule or algorithm for statistical inference. In practice it is the canons of statistical inference and all of the formal and informal rules that are part of knowing how to apply them that relate probabilities to statistics.

16 This is an ill-defined, and somewhat open-ended defeasibility condition that is nevertheless workable in practice. We usually recognize selection procedures that defeat the expectation that the statistics in the sample will reflect those in the target, even if we can't provide an explicit criterion that covers all cases.

17 See Ismael (2008, 2015b), Hall (1994) Lewis (1994), Pettigrew (2015) for disputes about the proper formal expression of the principle. The discussion in this section draws on Ismael (2008, 2011).

18 This substitutes Lewis' informal characterization of inadmissible information (as magical information from crystal balls and the like), with a physical characterization of inadmissible information as information about future events. In a relativistic setting, this is generalized so that inadmissible information is information not drawn from events in an agent's past light cone (see Ismael, 2009; Healey, 2016, 2017). Note that the generalization addresses one of Hall's objections to retaining the Lewisian view of chance as indexed to a time. Hall's framework for representing chances, which employs ur-functions, makes it easier to assimilate chance to familiar Bayesian reasoning, but it obscures the epistemic role of chance and makes it less easy to see how chances are distributed across space-time and related to the manifold of categorical fact. For that reason, I have retained the Lewisian framework.

19 Hoefer (2007) offers a Humean view of general probabilities. He calls them 'chances,' but he is using 'chance' to refer to any form of objective probability, whereas I reserve the term for the single-case probabilities that play the role of chance in PP. Since every particular event either happens or does not, it is at the level of general probability that we can match theoretical predictions to observed frequencies. The probabilistic predictions of the theory meet the observed frequencies in typical ensembles of the relevant reference class, but there is a loose fit between the general probabilities and observed frequencies for two reasons: (i) general probabilities bear a probabilistic relationship to frequencies in typical ensembles captured by the law of large numbers; a failure to match the observed frequencies can lower the likelihood of a theory, but never directly refute it, and (ii) we can have good systematic reasons for denying the assignment of a probability or assigning a probability different from an observed relative frequency.

20 This is a reference to the complicated microstructure that a reference class has to have to support the attribution of a probability. Von Mises (1957) applies probability to what he calls *collectives* – hypothetical infinite sequences of attributes (possible outcomes) of specified experiments that meet the following requirements. A *place-selection* is an effectively specifiable method of selecting indices of members of the sequence, such

that the selection or not of the index i depends at most on the first $i - 1$ attributes. There are two axioms:

Axiom of Convergence: the limiting relative frequency of any attribute exists.
Axiom of Randomness: the limiting relative frequency of each attribute in a collective ω is the same in any infinite subsequence of ω which is determined by a place selection.

The probability of an attribute A, relative to a collective ω, is then defined as the limiting relative frequency of A in ω. This kind of restriction, or one rather like it is needed to support the epistemic role of probability guiding belief about typical, not systematically chosen, subselections from the reference class. The von Mises criterion is quite strict. We can be less strict about the kind of stability that is required. The conditions we impose will tell us the conditions under which we can assign a probability and the inferences are licensed by that assignment. The stricter the conditions on application, the stronger the inferences that are licensed.

21 Ismael (2008) and Sober (2010).
22 Hoefer (1997, 2007).
23 Ismael (2011).
24 The "=" here is an equality, rather than an identity.
25 Shalizi and Moore (2003). In that article, the authors are thinking primarily about coarse-grained states relevant to the prediction of a proscribed set of macrovariables, assuming an underlying microdynamics that fixes the evolution of the macrovariables. We extend it to fundamental theories by including all observables in the initial set of macrovariables.
26 In a world in which we have information from the future, the chance-credence link is broken and so the chances provide no guidance.
27 Proposed as a DEF in Ismael (2011).
28 Here is his phrasing when he says what he shares with Lewis: "Objective chances are not primitive modal facts, propensities, or powers, but rather facts entailed by the overall pattern of events and processes in the actual world." (Hoefer, 2007, p. 549) On the strong reading 'entailment' means reduction: objective chances just are distributed patterns in the manifold of fact. On a weaker reading, they are recommended credences, based on pattern of fact.
29 And notice here that making the distinction between the general probability and the single case probability explicit is essential to getting the logic of the inference right. One assigns a general probability that makes explicit the expectation that the regularity persists, and there is an additional step that transfers that general probability to the single case and offers it as a recommended credence.

References

Albert, David Z. (2000). *Time and chance.* Cambridge, MA: Harvard University Press.
Albert, David Z. (2015). *After physics.* Cambridge, MA: Harvard University Press.
Cartwright, Nancy. (1999). *The dappled world: a study of the boundaries of science.* Cambridge, UK: Cambridge University Press.
Cohen, Jonathan, & Callender, Craig. (2009). A better best system account of lawhood. *Philosophical Studies, 145*(1), 1–34.
Diaconis, Persi. (1998). A place for philosophy? The rise of modeling in statistical science. *Quarterly of Applied Mathematics, 56*(4), 797–805.

Emery, Nina. (2015). Chance, possibility, and explanation. *The British Journal for the Philosophy of Science, 66*(1), 95–120.

Emery, Nina. (2017). A naturalist's guide to objective chance. *Philosophy of Science, 84*(3), 480–499.

Frigg, Roman, & Hoefer, Carl. (2007). Probability in GRW theory. *Studies in History and Philosophy of Science Part B: Studies in History and Philosophy of Modern Physics, 38*(2), 371–389.

Hájek, Alan. (2012). Interpretations of probability. In Edward N. Zalta (Ed.), *The Stanford Encyclopedia of Philosophy (Winter Edition)*. URL=<http://plato.stanford.edu/archives/win2012/entries/probability-interpret/>

Hall, Ned. (2010). David Lewis's metaphysics. In Edward N. Zalta (Ed.), *The Stanford Encyclopedia of Philosophy (Winter Edition)*. URL=<https://plato.stanford.edu/archives/win2016/entries/lewis-metaphysics/>

Hall, Ned. (1994). Correcting the guide to objective chance. *Mind, 103*(412), 505–517.

Hall, Ned. (this volume). Chance and the great divide.

Healey, Richard. (2016). Local causality, probability and explanation. In Mary Bell & Shan Gao (Eds.), *Quantum nonlocality and reality: 50 years of Bell's theorem*. Cambridge, UK: Cambridge University Press.

Healey, Richard. (2017). *The quantum revolution in philosophy*. Oxford: Oxford University Press.

Hoefer, Carl. (1997). On Lewis's objective chance: Humean supervenience debugged. *Mind, 106*(422), 321–334.

Hoefer, Carl. (2007). The third way on objective probability: a sceptic's guide to objective chance. *Mind, 116*(463), 549–596.

Ismael, Jenann. (2008). Raid! The big, bad bug dissolved. *Noûs, 42*(2), 292–307.

Ismael, Jenann. (2009). Probability in deterministic physics. *Journal of Philosophy, 106*(2), 89–108.

Ismael, Jenann. (2011). A modest proposal about chance. *Journal of Philosophy, 108*(8), 416–442.

Ismael, Jenann. (2015a). How to be Humean. In Barry Loewer & Johnathan Schaffer (Eds.), *A companion to David Lewis* (pp. 188–205). Oxford: Wiley Blackwell.

Ismael, Jenann. (2015b). In defense of the chance-mixing principle: response to Pettigrew. *Noûs, 49*(1), 197–200.

Ismael, Jenann. (2017). An empiricist's guide to objective modality. In Matthew H. Slater & Zanja Yudell (Eds.), *Metaphysics and the philosophy of science: new essays*. New York: Oxford University Press.

Lewis, David. (1980). A subjectivist's guide to objective chance. In Richard C. Jeffrey (Ed.), *Studies in inductive logic and probability, Vol. II* (pp. 263–293). Berkeley, CA: University of Berkeley Press.

Lewis, David. (1994). Humean supervenience debugged. *Mind, 103*(412), 473–490.

Loewer, Barry. (2004). David Lewis's Humean theory of objective chance. *Philosophy of Science, 71*(5), 1115–1125.

Maudlin, Tim. (2007). *The metaphysics within physics*. Oxford: Oxford University Press.

Pearl, Judea. (2001). Bayesianism and causality, or, why I am only a half-Bayesian. In David Corfield & Jon Williamson (Eds.), *Foundations of Bayesianism (Applied Logic Series Vol. 24)* (pp. 19–36). Dordrecht, The Netherlands: Kluwer Academic Publishers.

Pettigrew, Richard. (2015). What chance-credence norms should not be. *Noûs, 49*(1), 177–196.

Shalizi, Cosma Rohilla, & Moore, Cristopher. (2003). What is a macrostate? Subjective observations and objective dynamics. [ArXiv preprint at cond-mat/0303625]

Sober, Eliot. (2010). Evolutionary theory and the reality of macro probabilities. In Ellery Eells & James H. Fetzer (Eds.), *The place of probability in science* (pp. 133–160). New York: Springer.

von Mises, Richard. (1957). *Probability, statistics, and truth (Revised English edition)*. (Hilda Geiringer, Ed.). New York: Macmillan.

Study Questions for Part V

1. What is the Humean conception of chance? What is the anti-Humean conception? Illustrate both on the T-juncture example.
2. What is the Principal Principle and what problem does it pose for Humeans?
3. What are general conditional probabilities, chances, and credences? What are the relationships between them?
4. What is the best system analysis? And what is the half-Humean interpretation of it?

Part VI

Are Sexes Natural Kinds?

11 Are Sexes Natural Kinds?

Muhammad Ali Khalidi

11.1 What Are Natural Kinds?

We classify biological organisms into many categories. Most obviously, we sort them into species (e.g. the common fruit fly, *Drosophila melanogaster*). We also group them into higher phylogenetic taxa, such as genera (e.g. *Drosophila*), families (e.g. Drosophilidae), orders (e.g. Diptera), classes (e.g. Insecta), and so on. There are plenty of other biological categories besides, which do not correspond either to species or to higher taxa, such as *herbivore* and *carnivore*, *viviparous* and *oviparous*, *larva, pupa*, and *imago, nocturnal* and *diurnal, altricial* and *precocial*, and *male* and *female*, among others. On what basis do we make these classifications? The short answer is similarity, but as many philosophers have pointed out, similarity is subjective and is not amenable to strict criteria (Quine, 1969). In some sense, everything is similar to everything else.

A more promising answer is that we categorize individuals on the basis of shared properties, though that answer is also not without its problems. One deep philosophical puzzle concerns what to count as a property. We can conjure up negative properties (e.g. *nongreen*), disjunctive properties (e.g. *green or blue*), and convoluted properties (e.g. *green if examined before the year 2100, blue otherwise*). But these do not seem like genuine properties, though it is a challenge to spell out what disqualifies them as serious contenders. Setting aside such philosophical puzzles about the nature of properties, we can say that observable or measurable properties of individuals are generally the basis for classifying them into categories, and that classification is indispensable for understanding the world, including the biological world. Without categories, we might conceive of the world only as a series of individuals and we would be hard pressed to make generalizations across individuals or propose systematic laws or theories.

Classification is central to many of our endeavors, including law, religion, and sports, but science is undeniably the enterprise with the most elaborate systems of categories, and most of them work impressively well for explaining and predicting various things. Scientists group things into a multitude of categories and these classification schemes or taxonomic systems serve as the basis for scientific theorizing. As philosophers have long pointed out, some of these classification schemes seem less "natural" than others, at least with hindsight. The nineteenth-century philosopher of science William Whewell noted that a classification of plants on the basis of the numbers of pistils and stamens in their flowers, was "artificial," not natural (1847, pp. 489–490). His reason appears to be that this classification does not reveal any other similarities among the plants and it does not enable us to make significant generalizations (e.g. *All plants with two stamens are…*). It is a classification that does not do any work for us and, as such, has been discarded as science has progressed. The classifications that remain tend to be those that can be used successfully in inductive inferences and enable us to explain many features of the individuals classified. The reason that they are so successful is that they are based on properties that reliably indicate the presence of numerous other properties. If I classify a plant as an *angiosperm* (or flowering plant), I can infer that its stamens will have two pairs of pollen sacs, that the seeds will be enclosed within an endosperm, which is a nutritive tissue that provides food for the embryo, and so on. These generalizations enable us to make reliable predictions, for example, that if a hitherto unidentified plant is an angiosperm, its seeds will be enclosed within an endosperm, among other things. That provides a stark contrast from a classification based on the number of pistils or stamens, which does not enable us to make any generalizations or inferences. That is why categorization into *angiosperms* and *gymnosperms* persists in scientific practice, while classification on the basis of the number of stamens does not. We could say, following Whewell, that the successful classification schemes are the "natural" ones and are likely to persist.

It would be good to pause here and consider whether this classification is not merely predictive but also genuinely explanatory. Suppose you ask for an explanation of the fact that some plants have seeds enclosed within an endosperm, and I respond by saying that there is a simple explanation, namely that they are angiosperms. Have I really explained that fact? There may be contexts in which that would be considered an explanation, but there are surely others in which it would be a cop-out. As an explanation, it does not seem to go nearly far enough. One thing missing from this explanation is an account of *why* these plants have this property. A more complete explanation would surely be that it is because they are descended from other plants that have this same property and that this property was selected for over a very long period of time because of its adaptive advantage to the plants, in accordance with Darwin's theory of natural selection.[1] According to that theory, heritable variations in individual organisms that offer an adaptive advantage will tend to be transmitted to the next generation and preserved. There are two features of this explanation that deserve expanding. The first is its reference to a history of descent. Because natural selection, and the theory of evolution more generally, explains current properties of individuals based on the properties of their precursors, we often classify individuals in biology on the basis of a common history. Members of any given species are all thought to be descended from the same ancestors, and that is what explains why they have so many shared properties. Moreover, properties in the biological realm tend to be shared for a reason, namely their causal efficacy in survival and reproduction. Thus, a fuller explanation in this case would refer to the fact that all angiosperms are descended from ancestors that had this property *and* that the property itself was presumably instrumental to the survival of those ancestors. Indeed, this is the basis of the classification of plants into *angiosperms* and *gymnosperms*, not merely their causal properties but their etiology or causal history. Bearing all that in mind, it does seem as though classification can be explanatory when we supply the entire basis for placing an individual into a particular category, not just the category label.

A question that often arises is, just because some classifications are useful for explaining and predicting phenomena, can we conclude that they correspond to real divisions in nature? Or to put it more philosophically, can we go from an *epistemological* distinction (explanatory and predictive category) to a *metaphysical* one (natural kind)? One thing to say in response is that if our scientific inquiries do not reveal which categories correspond to nature's own kinds, then it is not clear how else we would go about determining that. Even though our current scientific categories may not be the definitive ones, science seems to our best bet for isolating natural kinds. Nevertheless, some philosophers tend be more restrictive, distinguishing between those scientific categories that are merely useful and those that correspond to natural kinds. For them, a natural kind must have an essence or essential property, and categories that do not group things together on the basis of a shared essence do not correspond to natural kinds. But essences are not easy to come by in the natural world, especially in biology. Essences are supposed to be properties that are linked to kinds by way of necessary and sufficient conditions, but there are few if any such strict links in the biological realm. It was once thought that being a member of a species would be a matter of having a certain combination of genes, so that any individual who had those genes would be a member of some species S and no individual could be a member of species S unless it had that specific combination of genes. But things turn out to be far more complicated: there is no unique set of genes that is both necessary and sufficient for being a member of any particular species. Essences are also thought to be intrinsic, but at least some biological properties are relational or extrinsic.[2] Finally, essences are commonly posited to be microstructural, but many biological properties are not. So the essentialist approach is not promising, at least when it comes to many biological properties and kinds.

A more auspicious approach is the theory that natural kinds are *homeostatic property clusters* (HPC) (Boyd, 1989). The gist of this account is that properties cluster together reliably because they tend to be kept in a state of equilibrium (homeostasis) by a causal mechanism that ensures this. This theory accords better with some biological kinds, but it may also be too constraining in this context, since biological kinds are often in flux and do not persist in a state of equilibrium for long.[3] Moreover, there is not always a single causal mechanism that generates the properties associated with a given biological kind (as we already saw in the case of species not being identifiable with a unique genotype). However, on a more liberal account of what natural kinds are, some loose combinations of properties tend to cause the instantiation of other properties, and these property clusters, which remain relatively stable over time, are what we designate as natural kinds. Individuals that have these properties are categorized together, and the usefulness of the categories stems from the fact that the properties cluster together reliably, not by accident but due to causal connections between them. On a "simple causal theory" of natural kinds, they correspond to clusters of properties that are causally linked to other properties (Craver, 2009; Khalidi, 2013). Thus, when we observe that one individual has some subset of these properties we can infer that it is likely to have the others. In other words, the epistemic utility of the categories is grounded in the causal clustering of the properties. In what follows, I will be operating with this more permissive understanding of natural kinds, in order to examine the plausibility of regarding the sexes, *female* and *male*, as natural kinds.

As already seen, members of biological kinds are often grouped together due to a shared history of descent. This is true particularly of the higher taxa (e.g. genera, classes, orders, etc.). The order Insecta corresponds roughly to the common category *insect*, and every schoolchild knows that all insects have six legs, a three-part body (head, thorax, abdomen), compound eyes, and two antennae. But many biologists would say that they are classified together not just because of their shared causal properties, but primarily on the basis of having the same origin. Like many other biological kinds, they can be considered at once causal kinds, grouped together on the basis of shared properties, and etiological kinds, grouped together on the basis of a common causal history. Along these lines, Millikan has observed that many natural kinds in biology are "copied kinds"; members of these kinds share properties *because* they have been copied from a common template, and those properties are likely to be copied for a reason. Millikan (2005, pp. 307–308) associates three features with individual members of copied kinds: (1) all members have been produced from one another or from the same models; (2) members have been produced by, in, or in response to, the same ongoing historical environment (including other copied kinds); (3) some "function" is served by members of the kind, where "function" is roughly an effect raising the probability that its cause will be reproduced. She also indicates that (1) is the primary characteristic of copied kinds, while (2) and (3) support it. Many biological kinds can be considered copied kinds in this sense. For instance, members of a biological species are descendants of the same ancestral population, they are produced in response to the same or a similar environment, and the production of one member raises the probability of the production of others. Hence, they are individuated primarily by their etiology or causal history, but they also share many causal properties (morphology, behavior, and so on) as a result of that common history and the copying process that it involves. Copied kinds constitute a particular type of natural kind, which are individuated *both* by their synchronic causal properties and their causal history (etiology). Moreover, copied kinds are special because classification by causal properties and causal history tend to converge on the same individual members. (Sometimes individuals grouped together based on a shared causal history have few if any causal properties in common.)

11.2 A Sex Primer

One way of approaching the question as to whether sexes are natural kinds is by asking whether the categories *female* and *male* are on a par with highly efficacious biological categories like species categories (e.g. *Drosophila melanogaster*), or whether the distinction between the sexes is more like the "artificial" distinction between plants based on the number of stamens and pistils, which was mentioned in Section 11.1. To address this question, it will be necessary to look more closely at the classification of organisms into *female* and *male*. To simplify matters and for reasons that will be explained in Section 11.3, I will restrict the discussion to animals, setting aside the plant kingdom for the time being.

Many animal species exhibit *polymorphisms*, whereby there are several distinct types within the species. In many social insects like ants and bees, there are different castes, such as workers, soldiers, and queens, and indeed, in some species there are different types of workers (all of whom are female). These are perhaps the most striking polymorphisms in the animal kingdom, though there are numerous others. Some polymorphisms arise as a result of genetic factors but others are due to environmental ones, as when queen bees are differentiated from workers as a result of diet. The most familiar polymorphism is the dimorphism associated with sexual reproduction, which occurs in a wide variety of species across the animal kingdom. In English, the word "sex" can refer to the activity or process of reproduction as well as to the distinction among two types of organism, and while the two phenomena are closely linked, they need to be distinguished. The sex differences that are found within many species are related to the existence of sexual reproduction. Living things originally reproduced asexually, as many organisms still do (e.g. bacteria, many plants, and a few species of animals). Sexual reproduction was a later innovation and there are many theories as to why it arose. It has some apparent advantages, since mixing genetic material in every generation has the potential to yield new alleles (alternative forms of the same gene) and new combinations of alleles, and some of these may have adaptive value. But it also has some disadvantages, such as decreased efficiency owing to the fact that every act of reproduction takes two individuals, not just one (setting aside self-fertilizing hermaphrodites). It is still a subject of lively scientific debate as to why sexual reproduction arose and what its pros and cons are, but we can say something about sexual dimorphism without settling that question.[4]

In principle, the evolution of sexual reproduction does not necessitate the evolution of two and only two sexual morphs, and it does not entail that these two morphs correspond to what we commonly label "female" and "male." In many sexually reproducing species, there are hermaphroditic individuals, which can be considered both female and male (or neither), and some species are entirely hermaphroditic. In these species, there is only one morph, even though individuals reproduce sexually. Banana slugs are a hermaphroditic species in which any individual can mate with any other, and in the absence of a partner, individual slugs can self-fertilize. In other species, there are more than two morphs, with different sex roles when it comes to courtship, mating, reproduction, child-rearing, and related functions (about which more later). Having said that, most animal species that reproduce sexually are sexually dimorphic, comprising just two distinct morphs. In sexually dimorphic species, one of the two morphs is usually labeled "female" and the other "male." But given the vast diversity among living beings, what justifies the neat classification into *female* and *male* across species? Are there any recurrent objective differences that would warrant applying the same two labels in different species, and if so, what are the properties that constitute the basis for this distinction?

At first blush, one might think that females and males are distinguished primarily by their sexual organs, but there are some species in which the female has a "penis" or the male has a "vagina" (or as one might expect, both). In *Neotrogla* cave insects, the penis-like female organ, called a "gynosome," is used to suck out sperm and nutritious seminal fluids from the male, which has a vagina-like opening in its body. Closer to home, among mammals, the female spotted hyena has a structure like a penis, which is used in mating.

More fundamental than the sexual organs are the two cells (gametes) that combine together to form a fertilized cell (zygote) in sexually reproducing species. The vast majority of sexually reproducing species are *anisogamous*, which means that the gametes come in two unequal varieties and are produced by two types of organs called "gonads," which are the ovaries and testes. In these species, each of the two morphs has one type of gonad (ovaries or testes) and contributes a distinct type of gamete (ovum or sperm), which fuse together to produce a single zygote that combines genetic material from two parent organisms. In these species, one of the two types of gametes (ovum) is much larger than the other (sperm). While each gamete contains half the genetic material of the parent, the ovum is also resource-rich, providing nutrients for the embryo. Biologists label the morph that produces the larger gamete, *female*, and the morph that produces the smaller one, *male*. Thus, the basis of the female-male distinction in animals is the relative size of the gametes and the type of gonads that produce them. What remains to be seen is whether this distinction marks an important difference between two morphs across all these sexually reproducing species, and hence, whether it serves as the basis of a "natural" classification scheme.[5]

Building on the discussion in the previous section, one way to go about addressing this issue is by determining whether the properties associated with these morphs (namely, their relative gamete size) correlate with any other differences, and whether they enable us to explain and predict a number of other properties besides. At least since Darwin, there has been a common view among evolutionary biologists that the disparity in the relative size of the gametes in sexually reproducing species causes numerous other differences in females and males across many species. As Darwin emphasized, sexual selection is a variant of natural selection applied to traits that are particularly advantageous with regard to sexual reproduction (rather than mere survival).[6] In the *Descent of Man*, Darwin attempted to explain how gamete size might lead to different traits among males and females in many species. He reasoned that the larger gamete is not easily transportable, while the smaller gamete is more mobile (and indeed, often equipped with flagella that enable it to travel significant distances). The smaller gamete usually needs to migrate towards the larger one so that fertilization can take place. This means that, especially in the "lowly-organised animals," where fertilization is external to the body, it is advantageous to the male to "acquire the habit of approaching the female as closely as possible," and this habit has been preserved in other species, even where fertilization is internal (Darwin 1871/1981, p. 274). Therefore, according to Darwin, across a wide range of species, the small-gamete-producer (male) is more prone to seek out and pursue the large-gamete-producer (female), which is to say that there are different behavioral dispositions in males and females in these species.

Nowadays, this explanation is not widely accepted, since a male's approaching the ova is not the same as pursuing the female, and anyway, it is unlikely that such a trait would persist in taxa in which it is no longer clearly advantageous due to internal fertilization. More recent evolutionary biologists have focused, not so much on the differences in mobility among the differently sized gametes, but the varying amounts of effort that it takes to produce them. Since female gametes are large and rich in resources, they require a sizeable investment of time and energy to produce, while male gametes are small and do not require as much. Hence, males can afford to produce many gametes and attempt to fertilize as many female gametes as possible, whereas females can only produce a small number and, given their investment, need to ensure that they are fertilized by males with advantageous genes, resulting in a need for greater selectivity.[7] This difference is posited to lead to different pre-mating and post-mating behaviors, leading females to be more selective in their mating habits, males to compete more intensely with other males for reproductive opportunities, and females to invest more resources in parental care. Another difference that is sometimes cited is that in many species in which fertilization is internal rather than external, the female, which contributes the larger gamete, also carries the fertilized zygote, which means that females put even more investment in child-bearing and child-rearing, leading to differential mating strategies and parenting behavior. Also, in these species, there is uncertainty about paternity due to the fact that sperm from different males can compete internally to fertilize the ova. This means that females tend to have greater assurance than males as to which offspring are theirs, which implies that the former are more likely to invest in parental care than the latter.[8]

For many evolutionary biologists, the relative size of female and male gametes is the causal factor that accounts for different morphologies, mating strategies, parenting behaviors, and other properties of females and males. The size differential among the gametes means that the larger gamete tends to be less mobile, more scarce, and require greater resources than the smaller gamete. Moreover, in the case of internal fertilization, the zygote tends to remain within the parent that contributes the larger gamete, thus providing less assurance of paternity for the other parent. Males tend to acquire greater strength and natural weapons to compete against other males, as well as forms of ornamentation to attract females. For females, it is an advantage to develop morphological and behavioral characteristics that aid in protecting and caring for offspring. Though these properties are not universally found in all animal species, the causal link between relative gamete size and a number of other properties, which are in turn causally linked to yet other properties, accounts for observed morphological and behavioral differences among the morphs that contribute the two types of gametes.[9] If these theories are right, then the classification of organisms into *female* and *male* across a range of animal species really does serve as the basis for a "natural" rather than an "artificial" classification scheme (in Whewell's terms). Moreover, in accordance with the "simple causal theory" of natural kinds, possession of the different types of gonads that produce differently sized gametes is the causal property that brings about a number of other morphological and behavioral traits across a range of animal species. If these speculations are correct, the sexes can therefore be considered to be natural kinds.[10]

The sexes also seem to be etiological kinds and copied kinds (in Millikan's sense). Not only do females share causal properties with other females and males with other males, they are also reproduced from other females and males. However, they are not straightforward copied kinds, since it is obviously not the case that females only produce females and males only males (indeed, the whole point is that it takes one of each sex to produce one of either sex).[11] Nevertheless, all anisogamous animal species, those featuring dissimilar gametes, are likely descendants of the same anisogamous ancestors, so the presence of unequal gametes in a wide range of species has the same origin, as do the female and male morphs that produce them. Hence, female and male organisms are individuated across different species as a result of having arisen from the same causal process (partly due to a common environment and similar selective pressures).[12]

11.3 Challenges to Sexes as Natural Kinds

The case was made in the previous section that females and males can be considered to be natural kinds across animal species, both on the grounds that they share causal properties, as well as on the grounds that they are descended from the same ancestral types by means of a copying process. But this case is open to a number of objections, some of which will be considered briefly in this section.

The first objection points to the fact that there are some sexually reproducing species in which there are more than just two sexual morphs, including "masculine females" and "feminine males" (Roughgarden, 2013). Rather than sexual dimorphism, these species can be said to exhibit sexual polymorphism (as mentioned in Section 11.2), when it comes to morphology and to mating and parenting behaviors. But there are two responses that can be made when faced with such phenomena. One is to count them as exceptions that do not completely undermine the basis for considering females and males to be natural kinds. After all, generalizations in biology are rarely if ever ironclad, and so we should not expect generalizations about biological natural kinds like *female* and *male* to be universal among sexually reproducing animals. The other response is to say that species with multiple morphs do not constitute an exception at all, since the additional morphs appear to be sub-kinds of the kinds *female* and *male*. Indeed, labels like "masculine female" and "feminine male" tend to reinforce this conjecture, since they seem to assume that these morphs are sub-kinds of females and males respectively, rather than altogether different kinds. And just as the existence of the kind *Drosophila melanogaster* (species) does not undermine the existence of the kind *Drosophila* (genus), the existence of subordinate sexual kinds does not count against the existence of the superordinate kinds *female* and *male*.

Perhaps the most powerful objection to *female* and *male* being natural kinds is that the morphological and behavioral properties associated with gamete size are far more variable across animal species than was acknowledged in the previous section, thereby undermining the very basis of the distinction (cf. Dupré, 1986). According to some researchers, the causal pathways from relative gamete size to morphological differences, mating strategies, parental investment, and so on, are too diverse to ground any useful generalizations. For instance, though it is true that the morph producing smaller gametes tends to be more competitive in many species, this assumes a roughly equal sex ratio. But if males are, for some reason, much scarcer than females in a species or population, then they will tend to be less competitive than females. Similarly, even if females tend to invest more resources in parenting in many species, this obtains when offspring need care, but in precocial species, ones in which the young are relatively independent, females do not need to provide parental care. Once all such exceptions are taken into account, a critic might say, it will be clear that there is no clustering of properties associated with large and small gamete producers respectively, thereby undermining the presumed naturalness of the kinds, *female* and *male*.

The strength of this objection depends largely on the empirical evidence and the extent to which generalizations concerning females and males are manifested across a wide range of species. Biologists themselves seem divided over the scope of these generalizations and their utility. But the preponderance of evidence currently suggests that there are some important and interesting generalizations that hold across a wide variety of species, and moreover that some of these differences evolved independently in separate lineages (e.g. relative size of females and males, male ornamentation, courtship displays by males, and parental care by females). Moreover, even though these generalizations may be riddled with exceptions, that is because there are other causal factors that intervene, interacting with gamete size to issue in a range of effects. It is not that *female* and *male* do not represent important causal properties, but they interact with other causal factors in such ways as to lead to somewhat different outcomes in different species, populations, or even environmental circumstances. This means that some generalizations only hold within more restricted domains. Even those biologists who are somewhat skeptical that many generalizations can be made about females and males across all animal species acknowledge, for instance, that there is a "mammal model" of female-only care for offspring, and an "avian model" of biparental care (e.g. Kokko & Jennions, 2008). This suggests that at least some generalizations about females and males hold within taxa like mammals or birds, rather than across all taxa. This would not be a vindication of the claim that *female* and *male* are natural kinds, but it would give support to the view that *female-bird*, *male-mammal*, and so on, are. Thus, even if it turns out that *female* and *male* are not natural kinds across animal species, they may enter into subordinate natural kinds within certain taxa.[13]

There are other objections to the view that the sexes are natural kinds, which may be worth considering in brief. It was already mentioned that some individuals, and indeed entire species, are hermaphroditic, combining both female and male gonads and sex organs in the same organism. If we were to consider these organisms both female and male, it may be objected that *female* and *male* cannot be natural kinds, since one mark of a natural kind distinction is that no individual can simultaneously be a member of two natural kinds. But that does not necessarily rule them out as natural kinds, it would just mean that *female* and *male* are not mutually exclusive kinds. Moreover, it may be more plausible to consider such individuals to belong to neither kind rather than to both, which would defuse the objection entirely. A related objection would refer to the fact that in some species, sex determination results primarily from environmental rather than genetic factors, such as temperature or population density, meaning that the same individual can be sequentially female and male. (This is referred to as sequential hermaphroditism, by contrast with simultaneous hermaphroditism.) But that would not necessarily be a deal-breaker, since there are arguably other natural kinds such that individuals can belong sequentially to different kinds (e.g. atoms of chemical elements that undergo radioactive decay). It may also be objected that femaleness and maleness constitute endpoints of a spectrum rather than two sides of a dichotomy, since in some species (e.g. humans) there are individuals who fall along a continuum between female and male. "While male and female stand on the extreme ends of a continuum, there are many other bodies... that evidently mix together anatomical components conventionally attributed to both males and females" (Fausto-Sterling, 2000, 31). But this again would not disqualify them from being natural kinds, at least if we allow species to be natural kinds. Species themselves evolve from other species and there are often individuals intermediate between distinct species, at least around the time of speciation events. Finally, it may be said that *female* and *male* crosscut other natural kinds such as *species*, in the sense that individuals *a* and *b* can both be members of species *S*, and individuals *b* and *c* can both be members of the kind *female*, yet *a* and *c* may belong neither to the kind *S* nor the kind *female* (since *a* is male and *c* belongs to some other species *S**). But here again, it appears that many good candidates for natural kinds crosscut other natural kinds (Khalidi, 1998).

One could also question the claim that that *female* and *male* are natural kinds in the sense of etiological kinds or copied kinds. If it turns out that sexual reproduction or anisogamy evolved more than once in the animal kingdom, then the morphs *female* and *male* would not be copied kinds, since they would not all derive from the same templates. But even so, they may yet be natural kinds on the basis of shared causal properties, just as long as we allow analogous biological structures to belong to the same kinds, such as *wings* and *eyes*, which evolved multiple times in different animal lineages, not just homologous ones like *vertebrae* or *femurs*, which have the same origin (Weiskopf, 2011; Ereshefsky, 2012).

Finally, this discussion of the sexes has been restricted to animals, mainly because it is difficult to make the case that (some) sexually reproducing plants also come in two sexual morphs, *female* and *male*. While it is true that many plants are anisogamous and plant gametes also tend to occur in two varieties, one of which is larger than the other, individual plants tend to produce both gametes and generally self-fertilize as well as cross-fertilize. Even in species where individual "female" and "male" plants exist, the fact that they are not mobile in the ways that animals are, means that the causal consequences of anisogamy are very different. Without going into further detail, suffice it to say that there are few, if any, morphological (much less behavioral) commonalities among "female" plants and animals or among "male" plants and animals. Hence, the claim being defended here is merely that *female* and *male* constitute natural kinds in the animal kingdom, not among all living organisms.

11.4 Enough about Sex, What about Gender?

In the past several decades, it has become commonplace to distinguish sex from gender, and to say that *sex* is natural, while *gender* is social or "socially constructed." The first part of that claim would seem to accord with the argument being made here. But this claim is also sometimes taken to imply that *sex* is real whereas *gender* is fictitious or imaginary. However, a more plausible interpretation is that *sex* (the female–male distinction) is a kind that pertains to the biological domain, with causal links to biological phenomena, such as morphology, reproduction, hormones, secondary sexual characteristics, basic behavioral patterns, and so on. Meanwhile, *gender* (the feminine–masculine or woman–man distinction) is a kind that belongs to the social domain, with primary reference to the human species. This means that gender categories feature mainly in explanations in the social sciences and the kinds *woman* and *man* pertain to those disciplines. That does not mean that gender is not real, just that its causes and effects relate to the social world and social phenomena, such as employment, discrimination, exploitation, and so on. This widespread viewpoint is lucidly expressed by Friedman (1996, p. 78):

> In the familiar terms of the sex/gender distinction, "sex" is the biologically given basis of sex identity and sexuality. Biological sex comprises external and internal genital anatomy, anatomically secondary sex-characteristics, and certain hormonal and chromosomal combinations. The words "female" and "male" identify persons in terms of their biologically sexed natures. Gender, by contrast, encompasses traits and behaviors which mark, and are traditionally thought to express, those biological givens in cultural practice. Gender includes psychological qualities, intellectual traits, social roles, grooming styles, and other modes of self-presentation. The words "feminine" and "masculine" identify persons in terms of their genders.

Like some other social kinds, gender appears to have an underlying biological basis (with which it is often confused). Though I have argued that *sex* is a real kind, *gender* depends in part on the *perception* of sex, so whether or not gender is real or not does not depend on whether sex is, since social reality is constituted in part by our perceptions. (To compare, many biologists do not consider *race* to be a real biological kind, yet that does not rule out the possibility that *race* is a real social kind, which may depend on a mistaken perception of biological difference.) Moreover, even though *gender* may be a real social kind in many societies in the present, due to a widespread perception that biological differences among females and males have social implications, it may cease to be one in the future as the social consequences of biological sex diminish. After all, humans are also categorized by biologists according to blood type, but this does not seem to have (and ought not to have) any social repercussions.

Some theorists have argued not only that *gender* is a social construction, but that *sex* is too (e.g. Fausto-Sterling, 2000). As I've already suggested, this is sometimes taken to mean that *sex* is fictitious or unreal, but that would be misguided, since it seems to assume that social phenomena are less real than biological ones.

However, it could more plausibly be taken to mean that sex, like gender, is primarily a *social* phenomenon and that we have misconstrued the nature of sex, mistakenly thinking that it pertains to biology when it doesn't. If that is right, then the argument that I have been making in this chapter, that *sex* is a biological natural kind, is wrong. Since I have already presented some evidence to the contrary, I will not recapitulate it here. Having said that, it may be true that we sometimes exaggerate the effects of sex and think that some properties are biologically based, when in fact they are a result of social processes. Just because *female* and *male* are real biological kinds that does not mean that they are associated with a number of wide-ranging and unchangeable psychological or behavioral properties. For example, it may be mistaken to think, as some researchers have, that there are cognitive differences associated with these kinds (cf. Fine, 2010). And even if there are such differences, they may be swamped by the effects of learning and culture to the point that they are negligible in the human species. Hence, we should take the claim that sexes are natural kinds in the animal kingdom with a grain of salt; it does not imply that the biological differences among female and male humans do and should have social consequences, even though they have had such consequences in most societies for all of recorded history.*

Notes

1 This is not meant to imply that all persistent biological traits are adaptive, since some traits are transmitted across generations for other reasons.
2 As we saw, members of a species are classified together based on their common origin, which is an extrinsic property. Some essentialists allow for extrinsic essences and argue that members of a species share a historical essence (e.g. Griffiths, 1999), while other essentialists deny this and maintain that members of a species share an intrinsic essence (e.g. Devitt, 2008).
3 For a critique of HPC theory as applied to species, see Ereshefsky & Matthen (2005).
4 Some biologists have argued that sexual reproduction is more advantageous in changing environmental conditions and less advantageous in relatively stable conditions (e.g. Otto 2008). There is also some uncertainty as to whether sexual reproduction evolved only once among living creatures or whether it arose more than once.
5 It is not entirely clear why the gametes evolved such different properties in the first place. But it might be adaptive to specialize, with one gamete containing nourishment for the embryo as well as genetic material, while the other just contains genetic material. A recent collection of articles on the subject states: "The evolution of anisogamy, one of the major evolutionary riddles to remain unsolved in the nineteenth and twentieth centuries, emerges into the twenty-first century as potent a mystery as ever" (Togashi & Cox, 2011, p. 4).
6 Prum (2017) defends the view, which he traces to Darwin's later work, that sexual selection is not just the "handmaiden" of natural selection, but a separate type of process, which depends largely on aesthetic appeal rather than adaptiveness.

* I am very grateful to Laura Franklin-Hall for encouraging me to write on this topic, discussing it with me on a number of occasions, and sharing her work and insights. I would also like to thank Marc Ereshefsky and Olivia Sultanescu for detailed comments on earlier versions of this chapter, which led to a number of improvements.

7   An influential version of this argument was given in Trivers (1972), relying partly on work by Bateman (1948). But the argument has been criticized on various counts, see e.g. Kokko & Jennions (2008).

8   The claim is not that organisms engage in these behaviors consciously, but that they acquire behavioral dispositions that tend to be transmitted genetically, epigenetically, and via learning mechanisms, because they are adaptive.

9   The causal links between gamete size and these other properties are not direct and proximal; rather, these properties come to be associated with each of the two morphs as a result of selection pressures over many generations. For example, members of the small-gamete-producing morph who have natural weapons tend to compete successfully for access to reproductive opportunities, which causes them to have more offspring, which in turn causes the trait to spread in that morph in the population. For a classic statement of the distinction between proximate and ultimate causes in a biological context, see Mayr (1961).

10 It may be tempting to think of *female* and *male* as HPC kinds rather than simple causal kinds, with a causal mechanism that keeps the respective clusters of properties in equilibrium. But though it used to be thought that there was a single master gene, SRY, on the Y chromosome that controlled gonadal differentiation in many mammals, it turns out that matters are far more complex (Roughgarden, 2013, pp. 197–199; Ainsworth, 2015, p. 298). Moreover, the genetic mechanisms are very different in many other animals, for example reptiles and birds.

11 Richardson (2010, p. 836) considers *females* and *males* to be "dyadic kinds" on the grounds that "sexes are not autonomous, individual classes, but interdependent, permanently coupled, interacting, binary subclasses of species…" In this respect they are different from other biological kinds such as species, yet she still considers them to be natural kinds.

12 Franklin-Hall (this volume) adopts a view of this kind.

13 This seems to be the view advocated by Dupré (1986), who makes a powerful case for not considering *female* and *male* to be natural kinds across animal species.

References

Ainsworth, Claire. (2015). Sex redefined. *Nature, 518*(7539), 288–291.

Bateman, Angus J. (1948). Intra-sexual selection in Drosophila. *Heredity, 2,* 349–368.

Boyd, R. (1989). What realism implies and what it does not. *Dialectica,* 43(1–2): 5–29.

Craver, Carl F. (2009). Mechanisms and natural kinds. *Philosophical Psychology, 22*(5), 575–594.

Darwin, Charles. (1981). *The descent of man and selection in relation to sex.* Princeton: Princeton University Press. [originally published at 1871]

Devitt, Michael. (2008). Resurrecting biological essentialism. *Philosophy of Science, 75,* 344–382.

Dupré, John. (1986). Sex, gender, and essence. *Midwest Studies in Philosophy, 11*(1), 441–457.

Ereshefsky, Marc. (2012). Homology thinking. *Biology & Philosophy, 27*(381–400).

Ereshefsky, Marc, & Matthen, Mohan. (2005). Taxonomy, polymorphism, and history: an introduction to population structure theory. *Philosophy of Science, 72*(1), 1–21.

Fausto-Sterling, Anne. (2000). *Sexing the body: gender politics and the construction of sexuality.* New York: Basic Books.

Fine, Cordelia. (2010). *Delusions of gender: how our minds, society, and neurosexism create difference.* (W. W. Norton, Ed.). New York.

Franklin-Hall, Laura R. (this volume). The animal sexes as explanatory kinds.

Friedman, Marilyn. (1996). The unholy alliance of sex and gender. *Metaphilosophy*, 27(1–2), 78–91.

Griffiths, Paul E. (1999). Squaring the circle: natural kinds with historical essences. In R. A. Wilson (Ed.), *Species: new interdisciplinary essays* (pp. 209–228). Cambridge, MA: MIT Press.

Khalidi, Muhammad A. (2013). *Natural categories and human kinds: classification in the natural and social sciences*. Cambridge: Cambridge University Press.

Khalidi, Muhammad A. (1998). Natural kinds and crosscutting categories. *The Journal of Philosophy*, 95(1), 33–50.

Kokko, Hanna, & Jennions, Michael D. (2008). Parental investment, sexual selection and sex ratios. *Journal of Evolutionary Biology*, 21(4), 919–948.

Mayr, Ernst. (1961). Cause and effect in biology. *Science*, 134(3489), 1501–1506.

Millikan, Ruth G. (2005). Why most concepts aren't categories. In H. Cohen & C. Lefebvre (Eds.), *Handbook of categorization in cognitive science* (pp. 305–315). Amsterdam: Elsevier.

Otto, Sarah P. (2008). Sexual reproduction and the evolution of sex. *Nature Education*, 1(1), 182–187.

Prum, Richard O. (2017). *The evolution of beauty: how Darwin's forgotten theory of mate choice shapes the animal world-and us*. New York: Doubleday.

Quine, Willard Van Orman. (1969). Natural kinds. In *Ontological relativity and other essays* (pp. 114–138). New York: Columbia University Press.

Richardson, Sarah S. (2010). Sexes, species, and genomes: why males and females are not like humans and chimpanzees. *Biology & Philosophy*, 25(5), 823–841.

Roughgarden, Joan. (2013). *Evolution's rainbow: diversity, gender, and sexuality in nature and people*. Berkeley: University of California Press.

Togashi, Tatsuya, & Cox, Paul Alan (Eds.). (2011). *The evolution of anisogamy: a fundamental phenomenon underlying sexual selection*. Cambridge: Cambridge University Press.

Trivers, Robert. (1972). Parental investment and sexual selection. In B. Campbell (Ed.), *Sexual selection and the descent of man* (pp. 136–179). Chicago: Aldine Publishing Company.

Weiskopf, Daniel A. (2011). The functional unity of special science kinds. *The British Journal for the Philosophy of Science*, 62(2), 233–258.

Whewel, William. (1847). *The philosophy of the inductive sciences (vol. 1)* (2nd ed.). London: John W. Parker.

12 The Animal Sexes as Historical Explanatory Kinds

Laura Franklin-Hall

Though biologists identify individuals as 'male' or 'female' across a broad range of animal species, the particular traits exhibited by males and females can vary tremendously. This diversity has led some to conclude that cross-animal sexes (males, or females, of whatever animal species) have "little or no explanatory power" (Dupré, 1986, p. 447) and, thus, are not *natural kinds* in any traditional sense. This chapter will explore considerations for and against this conclusion, ultimately arguing that, properly understood, the animal sexes are "historical explanatory kinds," groupings that can be scientifically significant even while their members differ radically in both their current properties and their particular histories. Whether this makes them full-fledged natural kinds is a question I take up at the very end.

12.1 The Animal Sexes and Natural Kinds

Almost everything about sex is controversial. But, among biologists at least, one thing is not: that males and females differ invariably and fundamentally in the relative size of the gametes they produce. Even Joan Roughgarden – as ardent a critic of traditional sex and gender binaries as any – underscores that "'male' means making small gametes, and 'female' means making large gametes. Period!" (2013, p. 23). This simple criterion for biological sex may be applied to any anisogamous species, i.e., any in which sexual reproduction involves the fusion of differently sized gametes (e.g., egg and sperm), each normally containing a half-complement of chromosomes. Anisogamy itself has evolved repeatedly across life's tree and is ubiquitous, though not strictly universal, among sexual organisms (Lehtonen & Parker, 2014).

Of course, that male and female differ in this way does not mean that every living thing is either male or female. In some species – commonly in plants, and frequently in fish – hermaphrodism is the rule, with each organism capable of producing gametes both large and small. Asexual species, the mostly single-celled isogamous lineages in which gametes are uniformly sized, and the rare protozoan group (e.g., *Chlamydomonas euchlora*) whose gametes come in more than two size variants, also lack distinct male and female sexes.

Yet in the "vast majority of animal species," from humans to horseshoe crabs, from bustards to beetles, almost all organisms are either male or female, but not both, a system known as gonochory (Fairbairn, 2013, p. 9).[1] Of course, differences between male and female often extend well beyond the reproductive cells, with the sexes varying in mass, anatomy, and behavior besides.The male blanket octopus, for instance, not only transfers sperm to its mate using a sex-specific detachable delivery arm called a hectocotylus, but he is also comparatively puny – 40,000 times less massive than the female.The female Anopheles mosquito likewise varies from conspecific males: she possesses special mouthparts that permit her to collect blood meals used to provision developing eggs.

The sex differences just mentioned are species-specific. But biologists as far back as Darwin have also identified coarser-grained patterns in the traits typical of males and females across the animal world. Male animals, it seems, are often more competitive in seeking mates, and sport the ornaments and armaments that aid in such competition.[2] Females, on the other hand, are commonly more discriminating in mating, and more commonly provide the bulk of species-typical parental care (Janicke et al., 2016).Theorists try to account for these and other trends by tracing them, in one way or another, to a difference of gamete size – just the difference that, biologists submit, defines the sexes themselves (Fromhage & Jennions, 2016; Lehtonen et al., 2016; Schärer et al., 2012).

That there exists a scientific consensus on this universal standard of *what it is* to be male or female, and that this difference is widely (though not universally) thought to be "the basis for other differences between males and females" (Hall & Halliday, 1998, p. 219), may appear to indicate that the *cross-animal sexes* – groups consisting of *all male animals*, and of *all female animals* – are deep and important ones that any student of the living world should recognize, that they are, in other words, *natural kinds*.

What are natural kinds? Many ways we group things simply reflect our own priorities. Pests, for example, is a category encompassing the many rather different animals that humans regard as a nuisance, from cane toads to termites and feral cats. However, to think that there are natural kinds – in the view of many philosophers, at least – is to think that some special ways of grouping things trace "the structure of the natural world rather than the interests and actions of human beings" (Bird & Tobin, 2017). Scientists, it would seem, aim to uncover the contours of these kinds and represent them in their classifications, as Mendeleev did with his periodic table of elements.

What exactly is it about natural kinds that sets them apart?The most discussed proposal is *kind essentialism*. On this view, natural kinds alone have an *intrinsic essence*: properties possessed internally by each kind-member that (1) are necessary and sufficient for membership and (2) account, at least in part, for other kind-typical features (for discussion, see Reydon, 2012, p. 219). For instance, all and only gold atoms have nuclei with 79 protons, which is responsible – in concert with the physical laws – for the characteristic conductivity, reflectivity, and boiling and melting points of all pure samples of gold.

If having an intrinsic essence is the mark of a natural kind, do the cross-animal sexes qualify? In a rare discussion of the question, John Dupré concludes in the negative, arguing that, though gamete size may provide a suitably strict standard for sex membership (thus satisfying condition (1) above), it is not appropriately *explanatory* of other sex-typical features (as required by (2) above) (Dupré, 1986, 1995). The essence of gold – having 79 protons – underpins, at least in part, the metal's attractive yellow luster; due to this nuclear constitution, free electrons form an 'electron sea' that readily reflects much of the visible spectrum, blue light aside. But that female orangutans make large gametes does not, in the same way, bring about their sex-typical parental behavior. Instead, gamete size and extended care for young (i.e., nursing through age 8 in orangutans) are, in the view of many biologists, products of a common cause: the female orangutan's genetic constitution. Nevertheless, the genetic constitution of the orangutan cannot be understood as the essence of female *animals*, the kind at issue here. While sex in orangutans and other mammals is set by the familiar X and Y chromosomes – with XX normally putting mammals on the path to becoming female, and XY on the path to developing as males – this sex determination system is parochial; no gene, chromosome, or combination thereof brings about female (or male) characteristics across the whole animal kingdom.

This explanatory shortcoming in sex's purported defining characteristic promises to illuminate another notable feature of the cross-animal sexes, one that arguably also distinguishes them from kinds like *gold*: that there are no substantive generalizations true of all – or even of nearly all – males, nor of females. For instance, though females more often provide parental care than do males across the animal kingdom, among bony fish in particular the pattern reverses: sole male care is more common than either sole female care or joint care. And though males are generally the more decorated sex, among some shorebirds the female is the more brightly plumed. Dupré suggests that this variability should be unexceptional, given the explanatory failure of gamete size. After all, on the essentialist picture, generalizations across kind-members were expected precisely because a kind's defining property brought about – and thereby explained – sundry other features of the individuals possessing them. When this relation is lacking, it is no surprise that strong generalizations are likewise absent.[3]

But if the cross-animal sexes are not essentialist kinds, what are they? After all, even if they don't satisfy essentialism's strictures, they seem rather unlike the purely conventional, human interest-based groupings, such as pests, with which natural kinds are often contrasted. This chapter attempts to answer that question by presenting a novel account of the animal sexes according to which they are historical kinds. I hope to show that by understanding the animal sexes in the way that I propose, we can make sense of the genuinely explanatory calling that the cross-animal sexes have in evolutionary biology, while at the same time appreciating the mesmeric diversity in male and female characteristics evident across the animal world.

My discussion proceeds as follows. First, Section 12.2 emphasizes known diversity in animal sex characteristics, a diversity that Section 12.3 argues is still consistent with the existence of genuine *trends* in the characteristics of male and female animals. After reviewing the best explanation for these trends in Section 12.4, Section 12.5 presents the chapter's central theoretical claim: that to be a male (or female) animal is to be one whose reproductive developmental process originates – in a way I will spell out – in that animal's earliest small (or large) gamete-producing animal ancestors. After arguing in its favor, I conclude by considering whether the animal sexes – so understood – are not merely *explanatory* but also genuinely *natural* kinds.

12.2 Diversity in the Cross-Animal Sexes

A first important fact about the cross-animal sexes – one apparent even when considering just those species in which the sexes are separate – is how heterogeneous each are. In particular, males differ from fellow males, and females from fellow females, not just in their species-typical characteristics (e.g., the male rainbow trout has fins, while the male mountain wren has wings) but also in the traits that are distinctive, within a given species, of males, or of females. The penis can illustrate. The possession of a penis may initially seem a promising candidate for a distinctively male attribute, one possessed by males, whatever their species, and absent in females. Yet not only are penile structures themselves extremely diverse – inflatable or stiffened by bone; single or paired; helical, boomerang, or forked – but males of many species, including most fish, birds, and amphibians, lack them entirely. And, in some cases, it is the female who is so endowed; in *Neotrogla*, a genus of Brazilian cave-dwelling bark lice, the female inserts her "penis" into the male's "vagina" to harvest his sperm (Yoshizawa et al., 2014).[4]

Male and female diversity is even more evident in behavioral traits, such as in reproductive and parenting activities. For instance, though some may presume that it is the female's exclusive role to care for young, sole female care is but one pattern of many – in fact, an uncommon one, given that, in most species, juveniles are left to fend for themselves. Even focusing on just those groups that do aid their offspring – by providing defense, nutrition, or instruction – single-handed caretaking can be found among males, as well as among females. In giant water bugs, for example, the female lays newly fertilized eggs atop the male's back, which he goes on to carry, protect, and aerate until hatching. Males that are similarly solicitous are also found among seahorses, who gestate embryos in a special-purpose abdominal patch or pouch, as well as among amphibians like the Darwin frog, whose male ferries his metamorphosing brood within his vocal sac.

Diversity even extends to the very factors that lead embryos to develop into males, or into females, in the first place. As mentioned above, no particular gene or chromosome triggers the development of either male or female phenotypes across animals. And in some cases the environment is the determining factor. The green spoonworm's sex, for instance, depends on features of its social environment: larvae develop as female if they settle on unoccupied ocean floor, but as male if they settle near a conspecific female.[5]

Given such examples, just what is it that animal males, and females, always have in common? That is, knowing nothing about a creature – except that it is an animal male, or an animal female – what can we infer about it, beyond one feature of its gametes? Perhaps *nothing*, as it turns out (Gorelick et al., 2016). From a creature's anatomy, to its behavior and genetics, biologists have uncovered no feature – save gamete size – perfectly distinctive of either sex. In fact, when it comes to how nongametic sex traits are distributed between males and females, almost *anything is possible.*

12.3 From Diversity to Trends

And yet, that anything is possible does not mean that all possibilities are equally probable. In this case in particular, that stronger conclusion is almost certainly false: for a variety of anatomical and behavioral traits, some are much more common among males than among females; for others, the reverse.

Let's return to the penis, understood broadly as any structure used to transfer gametes from one sex partner to another prior to fertilization. Though this apparatus has evolved repeatedly and independently across animals – derived, developmentally, from the anal fins of some fish, the limb pathway of amniotes, and the sensory organs of spiders (Brennan, 2016) – it is almost invariably, Brazilian cave-dwelling bark lice notwithstanding, a feature of the male animal alone (Yoshizawa et al., 2014, p. 1001). An equally strong trend, to be expected given that male and female genitals coevolved, concerns females: that they are much more likely than males to possess receptive genitals – from the ovipore of the insect to the triple vagina of the kangaroo – that collect, and regulate the use of, gametes of the other sex. Females are also more often the sex that gestates offspring, a pattern so striking that some accounts of pregnancy restrict it to females *by definition*, a constraint recently disputed by those studying the rare species in which males become gravid (Stölting and Wilson, 2007, p. 884).

A second class of sex-correlated traits comes from the more complex study of behavior. For instance, among animals, it is common for males to engage in more within-sex competition than conspecific females, including in direct contests between adults for access to mates (Andersson, 1994; Clutton-Brock, 2017). Females, on the other hand, are commonly more reproductively discriminating, that is, less likely than conspecific males to mate with available members of the other sex. A weaker – but still apparently real – sex-linked trend concerns parenting behavior: females, more often than do males, provide the majority of parental care (Clutton-Brock, 1991).

For a third type of sex-correlated trait, consider the developmental systems that build male and female phenotypes. Though what biologists call the *primary* sex-determination mechanisms – of which the mammalian XY/XX scheme is an instance – vary substantially across animals, aspects of the causal process activated by these mechanisms (like that causing gonadal differentiation) are similar across many animal species. In particular, the DM (*Doublesex* and *Mab-3*) family of transcription factors "appears to be directly involved in sexual development in all major animal groups" (Herpin & Schartl, 2015, p. 1265). The proteins produced by these genes contribute to the development or maintenance of male gonads in particular – and, by way of this, other aspects of male phenotype – in animals from flatworm to chicken, from mouse to flea. This regularity in the process of sex differentiation – in contrast with radical diversity in primary sex determination – has inspired the slogan "[m]asters change, [but] slaves remain" (Graham et al., 2003).

12.4 Explaining Trends

With a variety of sex-linked trends now in view, we can ask: what explains them? In particular, given that it is possible for animal males and females to differ so radically among themselves – as illustrated by Section 12.2 – why are there *any* patterns in male and female traits? After all, most biologists hold that members of each cross-animal sex must – as a matter of definition – have just a *single* feature in common: large gametes for females, small gametes for males. But why should an organism's gamete size mean *anything at all* for its other features?

Before exploring possible answers, it is worth considering an important, and potentially debunking, reply: that bias – most notably, our tendency to see the natural world through the lens of normative gender roles – has significantly distorted the study of sex differences and, to some degree, the conclusions drawn within it (Ah-King & Nylin, 2010, p. 234; Gowaty, 1997; Tang-Martínez, 2016). For instance, Darwin's (1871) description of male animals as generally "courageous and pugnacious" (ibid., p. 516) and females as "with rarest exceptions ... coy" (ibid., 222) seems to have been shaped by a Victorian model of behavior (Dewsbury, 2005, p. 835; Richards, 2017). More recently, phenomena contrary to our own social expectations – e.g., promiscuous females (Hrdy, 1980) and discriminating males (Edward & Chapman, 2011) – appear to have been recognized by the research community only very late in the course of inquiry and only after contentious debate (Hrdy, 1986).[6] And even presuming that these particular errors have been remedied (often following attention from feminists, e.g., Haraway, 1991; Gowaty, 2003), other aspects of our total picture of animal sex differences may well remain misshapen due to their prejudicial origins.

Granting the justice of these concerns, what bearing should they have on our understanding of the trends in male and female traits described in Section 12.3? Ought they, perhaps, be discarded along with Darwin's anthropomorphic language of the 'coy' and 'pugnacious'? Though caution is indeed called for, even those critics who are most attentive to the role of sex and gender bias in science do not generally go so far as this. Rather than questioning the *existence* of such trends, critics more often emphasize a cluster of problems with the way behavioral trends in particular have sometimes been understood or described. For instance, Ah-King and Ahnesjö (2013) argue against the common practice of grouping individual reproductive behaviors into broader "sex roles," with the "male sex role" involving heightened reproductive competition, minimal reproductive discrimination and parental care, and the "female sex role," the reverse (as in Barlow, 2005; Williams, 1966). Among other issues, such usage may indicate that sex characteristics come in just two cohesive packages, when the reality – as we've seen – is more complex. A second and related problem lies with the occasional suggestion that trends in sex differences might actually be *laws* (e.g., in Bateman 1948, p. 352), language that could mislead us into thinking that the generalizations have few, if any, exceptions. Along the same lines, critics take issue with claims that sex differences were generated "inevitably" by certain evolutionary processes or transitions, or that certain trends follow as a matter of "logic" (e.g., in Avise, 2013; Parker, 2014). Even if there is reason to think that these trends are not mere accidents, claims of "inevitability" are belied both by the intra-sex diversity already reviewed, and by our current understanding of the process that gave rise to animal sex differences in the first place.[7]

And just what is that process? This question brings us to the main task of this section: to explore explanations for trends in male and female traits across animals. At the heart of virtually all accounts is that males produce comparatively small gametes, and females large.[8] Of course, as reviewed above, gamete size does not do its work in the way the essentialist would envision, that is, via an immediate influence of gamete size, in each particular organism, on that organism's other traits. Instead, going explanations are uniformly evolutionary, with theorists positing that ancestral organisms with smaller gametes experienced different selection pressures than those with larger gametes, with current patterns of sex difference the eventual result.

Yet two subtly, but importantly, different sorts of evolutionary explanations have been suggested, which I will call the 'direct anisogamy' and the 'original anisogamy' accounts. And though theorists seem to increasingly advocate a version of the original anisogamy story, it is worth describing the direct account first, both given its continuing influence and for the sake of contrast.

In particular, in a paper foundational to the modern discussion of the evolution of sex difference, Bateman suggested that widespread differences in male and female mating behaviors could be explained by the fact "that females produce much fewer gametes than males" (1948, pp. 364–5), a difference that he said follows from the fact that smaller gametes are energetically cheaper than larger ones. The resulting mismatch in gamete number, as Bateman saw it, had the consequence that reproductive success is, for males, roughly proportional to their number of mating partners, but not so for females. After all, all of a female's gametes can be fertilized by a single male, while a male will have sperm enough for the eggs of many females. Bateman argued that, given this difference, it pays, from an evolutionary point of view, for males to vie for access to as many females as possible. This has led males to develop, among other traits, an "undiscriminating eagerness," and females – to maximize the quality of their more strictly limited offspring – a "discriminating passivity" (ibid., p. 365).

I have labelled this explanation 'direct' based on the role that gametic differences play in the selective story told. In particular, according to the explanation, a difference in gamete number between the sexes – thought to be a close consequence of anisogamy – *itself* makes a difference to which mating behaviors are most adaptive in extant organisms (and in their recent ancestors), thereby accounting, at least in part, for which behaviors are observed today.

Is this explanation cogent? And can it account, not only for the mating strategies found in particular populations (e.g., in the fruit flies that Bateman himself studied) but also for such behaviors across animal species – and thus for trends in such behaviors? Though it may succeed with the former, it appears to founder with the latter. The main problem is not the existence of so many species in which the small gamete producer, i.e., the male, lacks the traits that the direct explanation seems to predict. This fact *is* important, since it indicates that gametic differences can, at best, have particular selective consequences *given certain background conditions*.

But the direct account's principal shortcoming is that it is not the correct evolutionary explanation of mating behaviors in so many of the species that *do* instantiate the focal trend, e.g., in which males *are* more reproductively competitive than females.

Rather than gamete size, or gamete number, *parental investment* is more often taken to be the explanatory linchpin in accounting for sex differences in mating behaviors within species.[9] As influentially defined by Trivers, parental investment is "investment by the parent in an individual offspring that increases the offspring's chance of surviving … at the cost of the parent's ability to invest in other offspring" (1972, p. 55). It includes the costs of gestation, offspring feeding, and guarding, as well as gamete production itself, though this last is sometimes but a small contributor that makes no difference to the overall balance of investment between males and females. For instance, though Mandrill females do indeed invest more in making each of their larger gametes (i.e., eggs) than do males in their smaller gametes (i.e., sperm), this difference is trivial in comparison with the orders-of-magnitude-larger commitment of female Mandrills via gestation and lactation.

Why is it that overall parental investment – or, more specifically, sex differences in parental investment – matters so much? To mention just one of many paths of influence: because members of the more investing sex are more often occupied with preparing for or assisting offspring, they are underrepresented among breeding-ready adults. As a result, it is comparatively easy for members of the more investing sex to find mates, and comparatively difficult for members of the less-investing sex. In such a situation, each member of the less-investing sex stands to gain more, reproductively, from devoting resources to traits that increase the chances of mate acquisition, thereby explaining the emergence – over evolutionary time   of such traits in members of that sex.[10]

Yet can the parental investment theory, as just sketched, account for trends in sex differences across animal species? Needless to say, it promises only to address behavioral variation between the sexes, leaving aside the other types of sex difference surveyed above. Even so restricted, parental investment theory is not equipped to explain behavioral differences on its own, for the simple reason that it says nothing about *which sex,* male or female, typically invests more. And facts about parental investment cannot be extracted from the presence of anisogamy itself, even presuming that gametic investment is generally proportional to gamete size.[11]

This limitation brings us finally to the 'original anisogamy' account, a label I use for a family of emerging explanations according to which sex differences across animals – in mating behaviors as well as in many other traits – are "ultimately rooted in anisogamy, [… but] not directly due to anisogamy" (Fromhage and Jennions, 2016, p. 5; see also Avise, 2013; Lehtonen & Parker, 2014; Lehtonen et al., 2016). Very schematically, these accounts suggest that sex differences in gamete size in ancient animal ancestors can explain contemporary sex differences by having brought about a series of intermediate evolutionary innovations, innovations that themselves more directly account for the constellation of current sex differences described above.[12]

How might ancient gametic sex differences have such interesting explanatory potential? By a process recently dubbed the 'sexual cascade.' This can be envisioned as a series of evolutionary transitions in which an earlier sex difference – in the first instance, a gamete size difference itself – "precedes and creates the selective pressure for the next" (Parker, 2014, p. 7), a process that is then iterated. For instance, heightened male reproductive competitiveness in a contemporary group might be immediately explained by increasingly female-biased parental invest- ment that evolved at an earlier time, which is itself explained – at least in part – by the evolution of female viviparity, and so on, back to the sex differences in gamete size that, according to biologists, define the sexes themselves.

To illustrate part of the process in more detail, consider the advent of internal fertilization, a reproductive system in which gametes fuse while inside the body of the male or female parent. This common adaptation has emerged in widely divergent animal lineages, in many cases from ancestral conditions in which male gametes joined with female gametes near – but still outside – the parents. Importantly, internal fertilization has not evolved to occur in males and females with equal frequency; it is vastly more common in females. And why? Here is one plausible account: though internal fertilization, in itself, can benefit both parents, as it gives resulting zygotes some protection from predation, *female* internal fertil- ization emerged more often because it is comparatively evolutionarily accessible; i.e., given the other sex differences preceding it, female internal fertilization can arise via a much more modest, and more probable, novel variation. In particular, prior to the evolution of internal fertilization, the male's, but not the female's, gametes were already actively mobile, a difference largely maintained to this day. Given this pre-existing fact about gamete mobility, a simple change – say, the female's withholding of her gametes at the exit of the reproductive tract – might have been enough to bring about the target trait. After all, even with the female's gametes at a slightly greater remove, the male's gametes may still have been able to reach them under their own steam. But the converse process, in which a female's gametes were somehow transported to within the male, would have required a more complex set of innovations, and partly for this reason, arose in only a handful of cases.[13]

From the transition just described, we can trace this sexual cascade both back- wards and forwards in time. Going backwards, the active mobility of male – but not female – gametes is usually explained by selective forces resulting from anis- ogamy itself (Lessells et al., 2009). Going forwards, female internal fertilization has been used to account, at least partly, for quite a number of the sex differences mentioned in Section 12.3, including the comparative frequency and complexity of the male intromittent organs (Eberhard, 1985), an increased uncertainty of male paternity (important because it can make male parental care less likely, see Queller, 1997), and the higher rates of female internal gestation and viviparity (Kalinka, 2015). And given that gestation often shifts the balance of parental investment strongly towards the gestator, the stage is then also set – via parental investment theory – to explain the heightened male reproductive competition found in many extant species (Avise, 2013).

Needless to say, just because this sequence unfolded countless times, in separate lineages, it does not follow that it unfolded in all lineages. Many present-day external fertilizers, for example, took a more direct path from differences in gametic investment to sex differences in reproductive behavior. Other lineages deviated from the cascade at a late point, leading to different suites of reproductive traits. For instance, females in some species evolved to prefer males who contributed hard-to-find resources, putting males in comparatively high demand and bringing about what is controversially called a "sex role reversal"; in other cases, ecological conditions have favored substantial – sometimes equal – investment by both parents.[14] That animal populations have taken variable paths, sometimes to other destinations, is only to be expected, and not just because a stochastic element exists in all evolutionary processes. More important for explaining the diversity we see today is that no trait is fitness-enhancing in itself, but only relative to an organism's other endowments and its environment – including the environment provided by the other sex. Because these have varied over the history of animal life, evolutionary outcomes have differed likewise.

And yet, without denying this diversity, the original anisogamy account suggests that animal lineages have traversed certain kinds of paths more commonly than others, a bias thought to be no accident: it resulted from the fact that the simple gamete size difference in the distant past of all animals increased the probability of certain further sex differences in a large range of background conditions (Fromhage & Jennions, 2016; Lehtonen et al., 2016). These differences, in turn, increased the probability of other differences still.

As I've reconstructed the original anisogamy account, gametic differences play a qualitatively different explanatory part than they did in the direct account. In particular, the original anisogamy explanation recognizes that sex differences in gamete size do not, at least in many cases, *currently* make a difference to the adaptiveness of alternative male and female behaviors.[15] What does make a difference are other sex differences that evolved in later parts of the cascade.[16] And yet because these further differences did not *replace* gamete size variation but were tacked on top of them, the original anisogamy view still promises to explain the trends in animal sex differences that are observed today.

12.5 The Animal Sexes as Historical Kinds

The original anisogamy approach is still under active development and is not universally accepted. The ancient origins of animal evolution remain shrouded in mystery, after all, and the sexual cascade picture is incomplete. But presuming its general contours are correct, what might this mean for the nature of the animal sexes themselves?

It suggests that the animal sexes may be rather different than the biologist's simple "definition" suggests, as I now aim to show. In particular, though an animal's sex can generally be *identified* by the size of its gametes, such differences – I submit – are *indicators only*, and fallible ones at that. Rather than being grounded in any shared current properties, I propose that the animal sexes be understood as *historical kinds*, and, in particular, as developmental historical kinds that tie an animal's sex to the origins of the developmental process by which its reproductive traits are formed. This historical feature, it turns out, is closely tied to – if not identical with – what explains trends in sex differences across animals. Consequently, if this proposal is right, the animal sexes will turn out to be explanatory kinds, though of a somewhat queer sort.

Individual species are the sorts of things most often understood as historical kinds in biology. But the species model cannot simply be applied to the animal sexes: we cannot say, for instance, that male animals are members of different *reproductive lineages* than female animals, a difference in virtue of which they constitute different historical kinds. Naturally, males and females are jointly required for reproduction, and thus males and females in any population collectively possess the same causal and genealogical antecedents; in the case of different-sexed full siblings, they have fully identical ancestors.

Given this obvious fact, in what sense could 'animal male' and 'animal female' be historical kinds – i.e., those grounded in shared relationships to the *past*, rather than in shared *current* properties? To see, start by conceiving of the sexes, within a species, as alternative developmental outcomes of the same basic biological resources – of largely identical and "fundamentally bipotent" genomes (Beukeboom & Perrin, 2014, p. 37).[17] Though different species exploit different "switches" to direct individuals down one or the other of these developmental paths, animals with separate sexes appear to enjoy the same two-sex system, in the following sense: this system derives from a common evolutionary source, the mechanisms that underpinned bipotency in the common ancestors of all animals.[18] Those ancestors, it turns out, were already anisogamous, and able to develop into two reproductive variants (Cunningham et al., 2017). And this capacity, it seems, has persisted to this day, through locally gradual – and ultimately radical – change in its molecular underpinnings.

Not only does the bipotent system have a common origin across animals, but so do the particular settings of that system – the particular developmental paths that generate the adults we call 'male' and 'female.' The best evidence for this is the existence of a few conserved features of animal male and female reproductive developmental systems, particularly in the genetic networks critical for gonadal differentiation.[19] Still, such conservation is not required for male-specific developmental systems, and female-specific systems, to share such origins. What *is* required, as I conceive of it, is that a *sex development lineage* links reproductive developmental processes – those that bring about traits of specific relevance to reproduction, from gonads to genitals to mating behaviors – activated in present-day males, and females, to those in males and females (respectively), in the founding animal population.

To grasp the idea of a sex development lineage, consider the two reproductive types – one producing smaller gametes, the other larger – in that earliest animal population, a group of coral-like organisms living about 700 million years ago. At that early point, sex differences were likely very minor, with males and females (characterized by gamete size) differing only in gametic features, and perhaps in the particulars of the gonadal tissues that produced them. As this population evolved, and repeatedly split, sex-specific traits – and the developmental processes that generate them – gradually changed. Yet through such change, parents and their same-sex descendants would have exploited reproductive developmental systems similar enough to be considered variants of the *same system*. The sex developmental lineage is defined as the extension of these local relationships in reproductive developmental systems from generation to generation. Based on the presence of a few conserved features across animal males and animal females – most notably, in the process of gonadal differentiation itself – it appears that such lineages link the developmental systems found in the earliest large- or small-gamete-producing animal ancestors, with processes instantiated in male and female animals today.

In making the case for a common origin in animal male and female reproductive developmental systems, I just implied that an animal's sex is set by some intrinsic feature – presumably the size of its gametes – which the sex developmental lineage happens to track. But what if we kick away the gametic ladder that we climbed to get here, and understand an animal's sex *in terms of* the history of the developmental system that produced it? This, in effect, is my positive proposal. An animal's sex, I want to suggest, is set by its reproductive developmental system, and in particular by whether that system is a variant – as determined by its sex developmental lineage – of the developmental process at work in its earliest male, or female, animal ancestors.[20] Putting things somewhat more carefully, my idea is this: an animal is *male* (or *female*) just in case its reproductive traits came about by way of developmental processes linked via a sex development lineage to the developmental processes responsible for reproductive features in that animal's earliest *small-gamete-* (or *large-gamete*)-producing animal ancestors. It is only those earliest ancestors whose sex, male or female, was set by gamete size directly.

To illustrate, consider two male animals: a Rhinoceros beetle sporting offensive armaments, and a Titi monkey gently ferrying newborns on his back. In virtue of what are they both male animals? The orthodox, gamete-based answer, is that they are male just because they produce comparatively small gametes. My alternative historical proposal is this: they are male in virtue of the fact that the developmental processes that brought about their reproductive traits are variants of the developmental processes activated in each of their earliest small-gamete-producing animal ancestors. And to be a variant of such an early process is to be part of the same sex developmental lineage.

This historical account can make better sense of biological categorization and explanatory practices, I believe, than can the orthodox view (i.e., the view on which males are just those organisms that produce comparatively small gametes, and females comparatively large ones). But in making this case, let me acknowledge that some of the theoretical presuppositions on which I rely cannot all be defended here. Moreover, the viability of the entire picture depends on the empirical claims discussed above being in the main accurate. Given this, my primary aim can only be to make the historical approach to the sexes a live option, rather than to close the case in its favor.

A first consideration supporting the historical account of the sexes is the simplest: that, like other historical accounts of kinds, such as the now-popular genealogical approach to species, it makes immediate sense of *diversity* in the current characteristics of kind-members – in this case, in male and female traits across the animal kingdom. Capturing such diversity is something that any historical account can do very naturally, in virtue of the scope that an extended causal chain gives to the play of course-changing contingencies, yielding kind-members with radically different characteristics. In the case of the sexes in particular, much time – perhaps as much as 700 million years – has passed since the developmental origin that all females (and males) appear to share.

My proposal's developmental element also contributes, in a different way, to its openness to intra-*species* diversity in male traits, and in female traits. For example, consider an individual who has, in the main, come to be via the 'female' developmental pathway. (This pathway, of course, will be female in virtue of its history, albeit a history that will be largely the same across members of the species.) And yet, for various reasons – say, due to a congenital anomaly or inherited condition that interfered narrowly with gamete production – such an individual might not produce sex-typical gametes. Yet, on the view under consideration, this will not undermine that animal's status as a full-fledged *female,* in contrast to the orthodox anisogamy account with its singularly gametic focus.[2122]

Consider, next, the flip side of the coin: coarse-grained trends in sex differences – and in male and female traits – across the animal kingdom. In accounting for such trends, I submit that the historical account has a direct explanatory payoff. To see this, first recall the original anisogamy explanation from Section 12.4: that though current gamete size does not directly explain trends in male and female traits, the presence of ancient gamete size differences in animal ancestors does, via the sexual cascade. On my construal, this explanation has three claims at its heart: (1) that ancient animals were anisogamous, and thus came in small-gamete- and large-gamete-producing variants, variants that differed by having activated sex-specific developmental pathways; (2) that this two-variant system persisted through change; and (3) that – as described by the sexual cascade – selection pressures biased evolutionary change in the underpinning of sex-specific developmental pathways towards some patterns of sex difference and not others.

As the reader may note, I've just made explicit a critical part of the full original anisogamy explanation, one that I earlier left unstated: that differences between male and female animals are invariably underpinned by sex-specific developmental pathways. Not only is this feature central to our present understanding of the evolutionary process – a point emphasized by recent work in 'evo-devo,' the evolutionary study of development – but it is also the key to how the historical account of the sexes works its explanatory magic. Let me explain.

To be a female (or male) animal, I've proposed, is to be one whose reproductive traits are the product of a developmental pathway that is a variant of, and successor to, the pathway activated in early large-gamete producers (or small). What this proposal does, in effect, is to pack into the 'key property' of animal males, and of animal females, two key elements of the original anisogamy explanation – sex-limited versions of (1) and (2) above. By requiring, for an animal to be one sex or the other, that these states of affairs hold, an animal's sex promises to explain, via the sexual cascade, the likelihood that it possess other sex-linked traits, such as those described in Section 12.3. For instance, partly by virtue of the origin of its developmental system, an animal female will have a relatively low probability of reproductive ornaments and armaments but a higher probability of having receptive genitals. Such probabilities, it is worth noting, are explanatory, and not merely predictive: given the key property of female animals, the sexual cascade explains their likelihood of having certain traits; it does not simply predict that they likely will. In consequence, I submit that, if the animal sexes are as I propose, they are explanatory kinds.[23]

We come now to a final consideration in favor of my proposal. On the orthodox account, it is true "by definition" that to be male is to produce small gametes, and to be female, to produce large ones. This, I said, is the one uncontroversial element of the biologist's understanding of sex. Yet, ironically, it is just this idea that I have rejected. Is this not fatal to my analysis? After all, are not biologists the experts on the contents of their categories?

Though I see the results of the biological sciences as nearly definitive in revealing the kinds pertaining to life, the explicit definitions that biologists offer of the sexes are not definitive in themselves. When trying to understand the nature of a kind, the philosopher of science should look to the role that a kind plays in the work of scientists and in the world itself.[24] It is when we take this more general view that we find reasons to favor my account. Still, it would be awkward if such a project revealed kinds radically different from those that scientists describe. But my proposal does not do this, due to the fact that, not long after aniosogamy evolved, gamete size became an entrenched output of the developmental processes associated with each sex, even as other aspects of that process varied. In consequence, the kinds defined by the biologist's more straightforward criterion, and the animal sexes as understood on my view, tend to extensionally overlap.

But this may not always be true, and when the orthodox and historical accounts threaten to come apart, it arguably puts the historical account in the better light. Consider sex categorization in *Drosophila bifurca*, a fruit fly with giant (6-cm-long!) sperm. Though *D. bifurca* eggs are still, by mass, larger than sperm, by other measures, length included, such sperm exceed the female's eggs in size, and are nearly equinumerous with them. Nonetheless, biologists seem not to have given the least thought to questioning whether the sperm-producers are 'male' and the egg-producers, 'female,' though this would be something to consider on the orthodox account of sex.[25] By contrast, leaving this unquestioned, as biologists do, is just what one would expect on the account I favor. After all, giant sperm have been produced via a gradual evolutionary process, continuous with the developmental process at play in the small-gamete-producing ancestors throughout the whole animal kingdom.

Of course, since even the *Drosophila*'s giant sperm remain smaller than eggs *by mass* – arguably the size comparison that matters most – this observation is hardly definitive. More telling would be a study of the categorization practices of biologists, one that asked researchers how they would sex-categorize *D. bifurca* – as well as other animal groups already enjoying 'giant sperm' (Vielle et al., 2016) – in the counterfactual scenario in which sperm evolved to be more massive than eggs. For this, though, we must wait for another day.

12.6 The Animal Sexes as Natural Kinds?

If the animal sexes are indeed explanatory kinds, and if their nature is as I describe, does this mean they are *natural kinds* as well? The answer is not straightforward, due partly to persisting controversy over what it takes for a kind to be *natural*. Given the wide range of traits possessed by males and females, the sexes will certainly not be natural kinds according to any strict view like traditional essentialism. On the other hand, they may qualify if you accept a highly permissive view, like Dupré's (1995) own "promiscuous realism," on which any discontinuity in nature, however fleeting or trivial, may be enough for this status. But these views, each motivated by certain metaphysical concerns, are positioned at the extremes. What would an account say whose aim was not to apply some prior metaphysical criterion to science but to reconstruct the principles driving scientists' *own* division between the natural kinds – those, roughly speaking, of central and lasting scientific significance – and more adventitious groupings?

In prior work, I present one such view. On the Categorical Bottleneck Account (Franklin-Hall, 2015a), the natural kinds are those categories that well-serve both our actual epistemic purposes – such as those of prediction and explanation – and those of a large range of inquirers relatively like us. Kinds may be capable of doing this in a variety of ways, two of which are particularly straightforward. First, they might possess a set of relatively fine-grained properties that invariably run in tandem, a trait that makes them predictively useful for myriad purposes. Second, they might have broad explanatory potential (something distinct from – though often associated with – having a predictive role).

How do the animal sexes perform by such a measure? On the one hand, despite real trends in inter-sex differences, the animal sexes remain – in many if not most respects – highly heterogeneous. This significantly limits their predictive usefulness. It is for their explanatory calling that the animal sexes are, I think, of greatest interest. Still, on the Categorical Bottleneck view, what's critical to natural kind status isn't just having *some* actual explanatory or predictive use for us, whatever our idiosyncratic interests happen to be, but to have such use for many inquirers relatively like us. Among other virtues this sets apart as widely important kinds (e.g., the chemical elements, and the biological species) from more special-purpose, though still scientifically respectable, groupings (e.g., benthic organisms). With this more demanding standard in mind, the case that the animal sexes are natural kinds seems, to my mind, uncertain. After all, though an inquirer aiming to account for broad and loose trends in sex characteristics will be well-served by the animal sex categories, an inquirer captivated only by sex *diversity* – in making sense of the fascinatingly unique combination of sex characteristics found in each of the world's 2 million animal species – may find that the cross-animal sexes serve no live predictive or explanatory ends. If so, we would end up with a rather interesting result: animal sexes as explanatory, but not natural, kinds.

Notes

1 Even in gonochoric species, some organisms may produce both (or neither) large and small gametes, but these individuals are rare, and this property is not typically understood as an adaptation. This contrasts with species in which hermaphroditism is species-typical, as is the case for about 5% of animal species (Jarne and Auld, 2006). Still, some may wonder how we should categorize those relatively few individuals of gonochoric species who are neither male nor female, according to the anisogamy standard. This subject is complex, and not one that I can pursue here. Suffice it to say that there is little or no pressure, from a scientific point-of-view, to construct a truly exhaustive division of animals into the traditional sex categories. Yet the question cannot be so easily put aside in human contexts: some legal and social systems *presume* that all people can be so categorized. For an influential exploration of this issue (though one not focused on limits of the anisogamy standard in particular), see Fausto-Sterling's (1993) "The Five Sexes."

2 There are just a few known animal species in which females possess more extensive armaments than males, among them the polyandrous Northern Jacana, whose females are equipped with wing spurs, and a few varieties of dung beetle, whose females have larger horns used in female-female contests for resources. By contrast, there are countless species, across many phyla, in which the males alone are weaponized. See Emlen (2014) for an accessible survey.

3 I have reviewed just one strand in Dupré's complex reasoning on the sex categories, and two further features of his view warrant note. First, having rejected both essentialism and the suggestion that the cross-animal sexes have any significant *explanatory* calling, Dupré does not conclude that they entirely lack a *scientific* role–though exploring any such role has not been his focus. Second, irrespective of their scientific interest, Dupré will remain content to consider the sexes *natural kinds*, though for him this means little; on his extremely permissive account, natural kinds extend well beyond the categories of science to Pontiacs, puff pastries, and much else besides.

4 See Schilthuizen (2015) for a recent overview of genital diversity and evolution, and Eberhard (1985) for the classic treatise on the topic.

5 See Bachtrog et al. (2016) for a review of sex determination mechanisms across animals. From a more philosophical point of view, my (2015a) explores a rationale for the scientific practice of holding that sex is 'determined' or 'caused' *either* by the environment *or* by genetics, given that appropriate genetic and environmental conditions are always jointly required for normal development.

6 There are even stronger grounds for concluding that research into *human* sex difference in particular–an important topic beyond the scope of this paper–has been influenced by intellectual and social preconceptions (for discussion, see Fausto-Sterling; 2000; Fine 2010).

7 Neither of the works just cited actually maintains that sex differences are generated in a fully inevitable way; their claims, instead, are that a regular pattern of difference evolve *when certain other conditions are fulfilled*. In this way, it is the *language* of inevitability that is the subject of possible critique.

8 The one possible exception to the consensus about the importance of gamete size in the explanation of sex differences is the 'time-in/time-out' explanation articulated Gowaty and Hubbell (2005), according to which mean behavioral differences between animal males and females result from universal behavioral dispositions–i.e., those present in both males and females–activated differently due to the different circumstances that particular organisms experience. Space does not permit me to explore this view

here, but a few comments are in order. First, it is notable that this proposal is directed exclusively at accounting for sex-linked behavioral differences. Given this, anisogamy would still need to be called on to account for anatomical patterns of sex difference. Second, it remains uncertain if the account's explanations of *behavioral* differences isn't still committed—albeit in a special way—to an explanatory role for anisogamy. After all, the systematic difference in circumstances between males and females—which are then said to activate different particular behaviors—will *itself* need an explanation. For elaboration of this second point, see Schärer et al. (2012).

9 Even Bateman recognized, in passing, the importance of pregnancy and other forms investment, though the strict letter of his proposed explanation is in the gametic terms that I rehearsed. Of importance here is not Bateman exegesis but rather having these two contrasting explanations before us, both of which have had currency among theorists.

10 Though I describe parental investment theory as the successor to the direct account, it too is an oversimplification. To account for the evolution of sexually competitive traits in particular species, biologists often appeal to two population-level features that are *influenced* by differences in parental investment but are distinct from it: the ratio of breeding-ready males to females (or a property closely related to this), and the Bateman Gradient (the slope of the relationship between reproductive success and mating success). For details, see Kokko et al. (2012).

11 It is not at all clear that differences in *gametic investment* between the sexes do directly track differences in investment in individual male and female gametes, given that males of most species produce a vastly larger number of gametes than do females, both over their reproductive lives and per offspring embryo generated.

12 Though based on an evident difference between the two kinds of explanations that biologists offer, the explicit contrast between direct and original explanatory accounts is my own. As an aside, some presentations of (what I call) the original anisogamy account (i.e., Fromhage & Jennions, 2016) are advertised as *vindicating* Bateman's original insight, by showing in a more complex way how anisogamy might have generated other sex differences.

13 In contrast to the account just given, it may be that internal fertilization, even in its initial stages, was actually more fitness-enhancing for females than for males, or that there were additional dynamics at work (see Lehtonen & Parker, 2019). What matters for my purposes is not that the explanation given, an elaboration of one that Darwin himself offered, be fully correct, but instead that the correct explanation take the same overall form

14 For reasons of space, I cannot pursue the place of our own species among these options, perhaps a foolish pursuit given the evident *variability* in reproductive behaviors across human populations. Still, given the substantial *paternal* care evident in so many human groups, and the presence of seemingly sexually selected traits of *female* humans (e.g., permanent breasts, absent in all non-human primates), it seems safe to say that we have not followed what we might call the 'simple cascade' trajectory, one leading to exclusively male sexual competition, and substantial parental investment only among females. And why have humans gone, at least somewhat, off-cascade? No doubt the answer is complex, but the explanation many well involve our long, and demanding, juvenile period. In particular, though our mammalian lineage bestowed upon us female internal fertilization, and substantially obligates female parental investment via gestation and nursing, the long-term dependency of human young, and the demands of raising such juveniles to be reproduction-ready adults, may have selected for substantial paternal care alongside maternal care, resulting in the complex situation we see before

us: in which some level of sexual competition, and non-trivial parental investment, is typical of human males *and* females both.

15 This isn't to deny cases in which current gamete size differences *do* make a difference to behavioral strategies–for instance, those bird species in which the female's egg is orders of magnitude larger than each of a male's individual sperm, and in which there are but minor further differences in parental investment between the sexes.

16 Just what does it mean for gametic differences to *not make a difference* in this way? For heuristic purpose, think of this as follows: in the counterfactual scenario in which gametic investment between the sexes was equalized–while all else about male and female physiology stayed the same–there would, in many species, be *no* change in the overall parental investment difference between the sexes, and no qualitative change in selection pressures on male and female reproductive behaviors.

17 Partly for this reason, Richardson suggests that the sexes be understood as "dynamic dyadic kinds" (2013, pp. 196–199). This interesting proposal, which I believe is compatible with my own, is aimed at characterizing the sexes *within* a species, rather than across species.

18 Though both hermaphroditic species, and those with alternative reproductive tactics, can be fit into this picture, the details are too complex to pursue here.

19 See Section 12.3 and Beukeboom and Perrin (2014, chapter 3).

20 This is closely related to the fact that the actual animal sexes–or, more narrowly, some of their gonadal and gametic traits–seem to be homologies, such that two animals are jointly male in the same way that our forearm bone, and one of the bat's wing bones, are both *humerus bones*. Though this may be correct as a point about the actual animal sexes, my account actually does not require a relationship so strong. In particular, for two animals to be male I do not require that their developmental system's actually originate in the *same* small- or large-gamete-producing ancestors; it would be sufficient if they traced to distinct early anisogamous animal populations. This feature of my proposal is not one I can motivate here, except to say that it is tied to the likely repeatability of the sexual cascade: even had there been distinct origins of anisogamy in different animal populations – or, even had anisogamy (along with other preconditions for the full cascade, such as mobility) evolved elsewhere in the universe – sexual cascade theory suggests that we should nevertheless expect the same series of intermediate and ultimate adaptations, and thus it would be worth categorizing such diverse creatures together. Given this, my actual view is that the animal sexes are what I call "type-historical kinds," but I leave this complication aside here.

21 One might wonder what happens, on my view, if an individual's development goes by way of an intermediate developmental pathway, one perhaps resembling those in its male and female parent equally. On my view, an individual with a genuinely intermediate developmental origin would be neither male nor female, but this presents no problem for the overall account. After all, as Khalidi (this volume) explains, the presence of intermediate or vague cases need not, in and of itself, undermine a kind's scientific credentials, nor its status as *natural*.

22 It is worth noting that the anisogamy standard can also be given a developmental gloss, making it more open to this particular kind of intra-species diversity in reproductive characteristics. See Byrne (2018) for details.

23 Some may object to the standard for *explanatory kinds* to which I am appealing, along the lines suggested by Devitt (2008). In particular, he maintains that no historical relation can be explanatory of the intrinsic properties of individual kind-members, nor, presumably, trends in those characteristics. To some degree, I think this is a matter of labeling, as even Devitt would want to distinguish between historical kinds with 'explanatory potential,' that is, whose key historical property made reference to

genuinely explanatory prior causes, and those that did not. All I need, in these terms, is that the animal sexes satisfy the former standard. Yet, since space doesn't permit me to pursue such niceties here, let me simply note that the main line of thinking in philosophy of biology is consistent with the label I use. Consider, for instance, Khalidi's (this volume) discussion of what it takes for a kind to be explanatory, both in the case of species and the sexes themselves. Thanks to Michael Strevens for pressing this objection.

24 This is consistent, for instance, with inquiries into the nature of 'fitness,' where philosophers of biology have often endorsed a dispositional view even when biologists themselves often 'define' fitness in terms of actual reproductive output, a choice plainly useful on practical grounds but which leaves fitness without an explanatory role.

25 See, for instance, Lüpold et al. (2016).

References

Ah-King, Malin, & Ahnesjö, Ingrid. (2013). The "sex role" concept: an overview and evaluation. *Evolutionary Biology, 40*(4), 461–470.

Ah-King, Malin, & Nylin, Sören. (2010). Sex in an evolutionary perspective: just another reaction norm. *Evolutionary Biology, 37*(4), 234–246.

Andersson, Malte. (1994). *Sexual selection.* Vol. 72. Princeton, NJ: Princeton University Press.

Avise, John. (2013). *Evolutionary perspectives on pregnancy.* New York: Columbia University Press.

Bachtrog, Doris., Mank, Judith. E., Peichel, Catherine L., et al. (2014). Sex determination: why so many ways of doing it. *PLoS Biology, 12*(7), e1001899.

Barlow, George W. (2005). How do we decide that a species is sex-role reversed? *The Quarterly Review of Biology, 80*(1), 28–35.

Bateman, Angus J. (1948). Intra-sexual selection in Drosophila. *Heredity, 2*(Pt. 3), 349–368.

Beukeboom, Leo W., & Nicolas Perrin. (2014). *The evolution of sex determination.* Oxford: Oxford University Press.

Bird, Alexander, & Tobin, Emma. (2017). Natural kinds. In Edward N. Zalta (Ed.), *The Stanford Encyclopedia of Philosophy (Spring 2017 Edition).* URL = <https://plato.stanford.edu/archives/spr2017/entries/natural-kinds/>

Brennan, Patricia L. R. (2016). Studying genital coevolution to understand intromittent organ morphology. *Integrative and Comparative Biology, 56*(4), 669–681.

Byrne, Alex. (2018). Is sex binary? *Arc Digital.* https://arcdigital.media/is-sex-binary-16bec97d161e

Clutton-Brock, Tim. (1991). *The evolution of parental care.* Princeton, NJ: Princeton University Press.

Clutton-Brock, Tim. (2017). Reproductive competition and sexual selection. *Philosophical Transactions of the Royal Society B: Biological Sciences,* 372.1729: 20160310.

Cunningham, John. A., Liu, Alexander. G., Bengtson, Stefan, et al. (2017). The origin of animals: can molecular clocks and the fossil record be reconciled? *BioEssays, 39*(1), 1–12.

Darwin, Charles. (1871). *The descent of man and selection in relation to sex* (Volume 1). London: Murray.

Devitt, Michael. (2008). Resurrecting biological essentialism. *Philosophy of Science, 75*(3), 344–382.

Dewsbury, Donald A. (2005). The Darwin–Bateman paradigm in historical context. *Integrative and Comparative Biology, 45*(5), 831–837.

Dupré, John. (1986). Sex, gender, and essence. *Midwest Studies in Philosophy, 11*(1), 441–457.

Dupré, John. (1995). *The disorder of things: metaphysical foundations of the disunity of science.* Cambridge, MA: Harvard University Press.

Eberhard, William G. (1985). *Sexual selection and animal genitalia.* Cambridge, MA: Harvard University Press.

Edward, Dominic A., & Chapman, Tracey. (2011). The evolution and significance of male mate choice. *Trends in Ecology & Evolution, 26*(12), 647–654.

Emlen, Douglas J. (2014). *Animal weapons: the evolution of battle.* New York: Henry Holt and Company.

Fairbairn, Daphne J. (2013). *Odd couples: extraordinary differences between the sexes in the animal kingdom.* Princeton, NJ: Princeton University Press.

Fausto-Sterling, Anne. (1993). The five sexes. *The Sciences, 33*(2), 20–24.

Fausto-Sterling, Anne. (2000). *Sexing the body: Gender politics and the construction of sexuality.* New York: Basic Books.

Fine, Cordelia. (2010). *Delusions of gender: How our minds, society, and neurosexism create difference.* New York: WW Norton & Company.

Franklin-Hall, Laura R. (2015a). Natural kinds as categorical bottlenecks. *Philosophical Studies, 172*(4), 925–948.

Franklin-Hall, Laura R. (2015b). Explaining causal selection with explanatory causal economy: biology and beyond. In Pierre-Alain Braillard and Christophe Malaterre (Eds.), *Explanation in Biology. Dordrecht: Springer.*

Fromhage, Lutz, & Jennions, Michael D. (2016). Coevolution of parental investment and sexually selected traits drives sex-role divergence. *Nature Communications, 7,* 12517.

Gorelick, Root, Carpinone, Jessica, & Derraugh, Lindsay Jackson. (2016). No universal differences between female and male eukaryotes: anisogamy and asymmetrical female meiosis. *Biological Journal of the Linnean Society, 120*(1), 1–21.

Gowaty, Patricia Adair. (1997). Sexual dialectics, sexual selection, and variation in reproductive behavior. *Feminism and Evolutionary Biology,* 351–384.

Gowaty, Patricia Adair. (2003). Sexual natures: How feminism changed evolutionary biology. *Signs: Journal of Women in Culture and Society, 28*(3): 901–921.

Gowaty, Patricia Adair, & Hubbell, Stephen P. (2005). Chance, time allocation, and the evolution of adaptively flexible sex role behavior. *Integrative and Comparative Biology, 45*(5), 931–944.

Gowaty, Patricia Adair, & Hubbell, Stephen P. (2009). Reproductive decisions under ecological constraints: it's about time. *Proceedings of the National Academy of Sciences, 106*(Supplement 1), 10017–10024.

Graham, Patricia, Penn, Jill KM, & Schedl, Paul. (2003). Masters change, slaves remain. *Bioessays, 25*(1), 1–4.

Hall, Marion, & Halliday, Tim (Eds.) (1998). *Behaviour and evolution,* Heidelberg: Springer Verlag in association with the Open University.

Haraway, Donna. (1991). *Simians, cyborgs, and women.* New York: Routledge.

Herpin, Amaury, & Schartl, Manfred. (2015). Plasticity of gene-regulatory networks controlling sex determination: of masters, slaves, usual suspects, newcomers, and usurpators. *EMBO Reports, 16*(10), 1260–1274.

Hrdy, Sarah Blaffer. (1980). *The langurs of Abu: female and male strategies of reproduction.* Cambridge, MA: Harvard University Press.

Hrdy, Sarah Blaffer. (1986). Empathy, polyandry, and the myth of the coy female. In Elliott Sober (Ed.), *Conceptual issues in evolutionary biology.* Cambridge, MA: MIT Press.

Janicke, Tim, Häderer, I. K., Lajeunesse, M. J., et al. (2016). Darwinian sex roles confirmed across the animal kingdom. *Science Advances, 2*(2), e1500983.

Jarne, Philippe, & Auld, Josh R. (2006). Animals mix it up too: the distribution of self-fertilization among hermaphroditic animals. *Evolution, 60*(9), 1816–1824.

Kalinka, Alex T. (2015). How did viviparity originate and evolve? Of conflict, co-option, and cryptic choice. *BioEssays, 37*(7), 721–731.

Khalidi, Muhammad Ali. (n.d.). Are sexes natural kinds?

Kokko, Hanna, Klug, Hope, & Jennions, Michael D. (2012). Unifying cornerstones of sexual selection: operational sex ratio, Bateman gradient and the scope for competitive investment. *Ecology Letters, 15*(11), 1340–1351.

Lehtonen, Jussi, & Parker, Geoff A. (2014). Gamete competition, gamete limitation, and the evolution of the two sexes. *Molecular Human Reproduction, 20*(12), 1161–1168.

Lehtonen, Jussi, and Geoff A. Parker. (2019). Evolution of the two sexes under internal fertilization and alternative evolutionary pathways. *The American Naturalist, 193*(5): 702–716.

Lehtonen, Jussi, Parker, Geoff A., & Schärer, Lukas. (2016). Why anisogamy drives ancestral sex roles. *Evolution, 70*(5), 1129–1135.

Lessells, C. K., Snook, Rhonda R., & Hosken, David J. (2009). The evolutionary origin and maintenance of sperm: selection for a small, motile gamete mating type. *Sperm Biology*, 43–67.

Lüpold, Stefan, Manier, M. K., Puniamoorthy, N., et al. (2016). How sexual selection can drive the evolution of costly sperm ornamentation. *Nature, 533*(7604), 535.

Parker, Geoff A. (2014). The sexual cascade and the rise of pre-ejaculatory (Darwinian) sexual selection, sex roles, and sexual conflict. *Cold Spring Harbor Perspectives in Biology, 6*(10), a017509.

Queller, David C. (1997). Why do females care more than males? *Proceedings of the Royal Society of London. Series B: Biological Sciences, 264*(1388), 1555–1557.

Reydon, Thomas A. C. (2012). Essentialism about kinds: an undead issue in the philosophies of physics and biology? *Probabilities, Laws, and Structures*, 217–230.

Richards, Evelleen. (2017). *Darwin and the making of sexual selection*. Chicago: University of Chicago Press.

Richardson, Sarah S. (2013). *Sex itself: The search for male and female in the human genome*. Chicago: University of Chicago Press.

Roughgarden, Joan. (2013). *Evolution's rainbow: diversity, gender, and sexuality in nature and people*. Oakland, CA: University of California Press.

Schärer, Lukas, Rowe, Locke, & Arnqvist, Göran. (2012). Anisogamy, chance and the evolution of sex roles. *Trends in Ecology & Evolution, 27*(5), 260–264.

Schilthuizen, Menno. (2015). *Nature's nether regions: what the sex lives of bugs, birds, and beasts tell us about evolution, biodiversity, and ourselves*. London: Penguin.

Stölting, Kai N., & Wilson, Anthony B. (2007). Male pregnancy in seahorses and pipefish: beyond the mammalian model. *BioEssays, 29*(9), 884–896.

Tang-Martínez, Zuleyma. (2016). Rethinking Bateman's principles: challenging persistent myths of sexually reluctant females and promiscuous males. *The Journal of Sex Research, 53*(4–5), 532–559.

Trivers, Robert. (1972). Parental investment and sexual selection. In B. Campbell (Ed.), *Sexual selection and the descent of man* (pp. 136–179). Chicago: Aldine Publishing Company.

Vielle, A., Callemeyn-Torre, N., Gimond, C., Poullet, N., Gray, J. C., Cutter, A. D., & Braendle, C. (2016). Convergent evolution of sperm gigantism and the developmental origins of sperm size variability in Caenorhabditis nematodes. *Evolution, 70*(11), 2485–2503.

Williams, George C. (1966). Natural selection, the costs of reproduction, and a refinement of Lack's principle. *The American Naturalist, 100*(916), 687–690.

Yoshizawa, Kazunori, Ferreira, R. L., Kamimura, Y., et al. (2014). Female penis, male vagina, and their correlated evolution in a cave insect. *Current Biology, 24*(9), 1006–1010.

Study Questions for Part VI

1. What are natural kinds according to Khalidi? Are "female" and "male" natural kinds on Khalidi's view? Why?
2. What are natural kinds and what are historical explanatory kinds according to Franklin-Hall? Are "female" and "male" historical explanatory kinds? Are they natural kinds?
3. Describe two objections to the claim that males and females can be considered to be natural kinds across animal species. How may one respond to these objections?

Suggestions for Further Reading

Part I: Are Boltzmann Brains Bad?

Albert, David Z. (2003). *Time and chance*. Harvard University Press.
Introduction to the philosophy of statistical mechanics and thermodynamics.

Albrecht, Andreas, & Sorbo, Lorenzo. (2004). Can the universe afford inflation? *Physical Review D*, 70(6). [arXiv:hep-th/0405270]
In this paper, physicists Albrecht and Sorbo develop the cosmic inflation theory. Their version of cosmic inflation theory is meant to overcome a number of challenges, including the Boltzmann Brain problem.

Carroll, Sean M. (2006). Boltzmann's anthropic brain. *Discover Magazine*
An accessible explanation of many physical concepts related to Boltzmann Brains, including: entropy, the second law of thermodynamics, entropy fluctuations, and how they are thought to be related to the existence of our universe and Boltzmann Brains.

Dyson, Lisa, Kleban, Matthew, & Susskind, Leonard. (2002). Disturbing implications of a cosmological constant. *Journal of High Energy Physics*, 10, 011. [arXiv:hep-th/0208013]
In this paper, physicists Dyson, Kleban, and Susskind argue that the Boltzmann Brain problem is an implication of assuming a cosmological constant (see especially section 6).

Elga, Adam. (2004). Defeating Dr. Evil with self-locating belief. *Philosophy and Phenomenological Research*, 69(2), 383–396.
Suppose one thinks that her subjective state is instantiated multiple times in the universe. How confident should she be that she is experiencing one instanton rather than another? Roughly, Elga argues that such agent ought to be uncertain which instantiation she is experiencing.

Hartle, James B., & Srednicki, Mark. (2007). Are we typical? *Physical Review D*, 75(12), 123523. [arXiv:0704.2630 [hep-th]]
This paper argues that there is no data to support the assumption that ordinary observers, like us, are typical in the universe. Therefore, cosmologies which imply that Boltzmann Brains are much more common than ordinary observes are consistent with the data we have. Kotzen's chapter in this volume responds to the argument made in this paper.

Page, Don N. (2007). Typicality defended. [ArXiv preprint at arXiv:0707.4169 [hep-th]]
Contra to Hartle and Srednicki, Page argues that there is data to support the assumption that ordinary observers, like us, are typical in the universe. Cosmologies which predict that Boltzmann Brains, rather than ordinary observers, are typical in the universe would be inconsistent with this data.

Part II: Does Mathematical Explanation Require Mathematical Truth?

Balaguer, Mark. (2001). *Platonism and anti-Platonism in mathematics*. New York: Oxford University Press.

Balaguer defends both mathematical Platonism and mathematical anti-Platonism (and especially fictionalism). He rejects prominent arguments made against these views, and argues that there is no fact of the matter on whether mathematical Platonism is true.

Colyvan, Mark. (2015). Indispensability arguments in the philosophy of mathematics. In Edward N. Zalta (Ed.), *The Stanford Encyclopedia of Philosophy (Spring 2015 Edition)*. URL =<https://plato.stanford.edu/archives/spr2015/entries/mathphil-indis/>

This paper introduces the indispensability argument attributed to Quine and Putnam. It includes an overview of the argument, discussion of important concepts, and overview of some objections to it.

Lange, Marc. (2017). *Because without cause: non-causal explanations in science and mathematics*. Oxford: Oxford University Press.

This book characterizes and analyzes noncausal explanations. The discussion includes explaining what mathematical explanations are, fleshing out distinctive kinds of noncausal explanations, and illustrating using examples from physics and mathematics.

Maddy, Penelope. (1997). *Naturalism in mathematics*. Oxford: Clarendon Press.

In this book, Maddy defines what naturalism in mathematics is, motivates it, and applies it to set theory. In addition, the book discusses indispensability argument in scientific and mathematical practice, and Quinean naturalism in particular.

McCain, Kevin, & Poston, Ted (Eds.). (2017). *Best explanations: new essays on inference to the best explanation*. Oxford: Oxford University Press.

This book is a collection of papers presenting different perspective on the inference to the best explanation. Among other topics, the essays in this book discuss the relation between inference to the best explanation to testimony, skepticism, and Bayesianism.

Part III: Does Quantum Mechanics Suggest Spacetime is Nonfundamental?

Ismael, Jenann, & Schaffer, Jonathan. (2016). Quantum holism: nonseparability as common ground. *Synthese*, 1–30.

This paper clarifies the meaning and rationale of quantum holism by articulating a strategy to explain the phenomenon of nonseparability.

Lange, Marc. (2002). *An introduction to the philosophy of physics*. Oxford: Blackwell.

This book introduces various physical concepts from a philosophical and historical perspective. Among other topics, the book covers quantum mechanics and in particular nonlocality and entanglement.

Leifer, Matthew Saul. (2014). Is the quantum state real? An extended review of ψψ-ontology theorems. *Quanta*, 67–155.

A comprehensive overview, aimed at a broad audience, of concepts, theorems, and literature related to wavefunctions realism.

Myrvold, Wayne C. (2015). What is a wavefunction? *Synthese*, *192*(10), 3247–3274.

This paper considers wavefunctions in light of the fact that quantum mechanics is just an approximation of relativistic quantum field theory. The paper concludes that wavefunctions are not part of the basic ontology of the world.

Myrvold, Wayne C. (2018). Philosophical issues in quantum theory. In Edward N. Zalta (Ed.), *The Stanford Encyclopedia of Philosophy (Fall 2018 Edition)*. URL = <https://plato.stanford.edu/archives/fall2018/entries/qt-issues/>

An overview of philosophical debates about quantum mechanics, including the measurement problem, entanglement, ontological issues, and quantum computing.

Ney, Alyssa, & Albert, David Z. (2013). *The wave function: essays on the metaphysics of quantum mechanics.* Oxford: Oxford University Press.

A collection of papers on wavefunction realism.

Part IV: Is Evolution Fundamental When It Comes to Defining Biological Ontology?

Folse III, Henri J., & Roughgarden, Joan. (2010). What is an individual organism? A multi-level selection perspective. *The Quarterly Review of Biology, 85*(4), 447–472.

This paper defines what an individual organism is using evolution theory, focusing on concepts of fitness and the concept of adaptive functional organization. Further, the paper defines a notion of evolutionary individual, such that species might count as evolutionary individuals.

Garvey, Brian. (2007). *Philosophy of Biology.* Stockfield, UK: Acumen.

This is a textbook covering various issues in philosophy of biology, including evolution theory, Darwinism, developmental biology, genes, species, laws, and ethical issues.

Lloyd, Elisabeth A. (1994). *The structure and confirmation of evolutionary theory.* Princeton, NJ: Princeton University Press.

In this book, Lloyd defends the scientific status of Darwinian evolution theory by comparing it to physical theories. The book includes a discussion of evolution theory meant for those without background in biology, alongside discussion which presupposes familiarity with evolution theory. It discusses what the structure of evolution theory is, what the units of selection are, and what confirmation is in the context of evolution theory.

Pradeu, Thomas. (2010). What is an organism? An immunological answer. *History and Philosophy of the Life Sciences, 32,* 247–267.

Pradeu discusses the notion of a biological individual from the point of view of immunology theory, and then links it to the evolutionary perspective on what a biological individual is.

Sober, Elliott (Ed.). (2006). *Conceptual issues in evolutionary biology* (3rd ed.). MIT Press.

A collection of papers on core debates in philosophy of evolution theory, including evolutionary ethics, social issues, laws in evolution theory, essentialism, and reductionism.

Part V: Is Chance Ontologically Fundamental?

Emery, Nina. (2017). A naturalist's guide to objective chance. *Philosophy of Science, 84*(3), 480–499.

This paper argues that we should believe that some probabilities exist, even if they are metaphysically weird and even if we can't give them a robust metaphysical analysis. We need them to be able to explain things in the world.

Hájek, Alan. (2012). Interpretations of probability. In Edward N. Zalta (Ed.), *The Stanford Encyclopedia of Philosophy (Winter Edition).* URL=<http://plato.stanford.edu/archives/win2012/entries/probability-interpret/>

A review of various concepts of probability. The review considers probability concepts from two perspective: how well they satisfy formal notions, and what roles they play in our conceptual apparatus. Among others, the review discusses Best-System probability concepts.

Lewis, David. (1994). Humean supervenience debugged. *Mind, 103*(412), 473–490.

In this paper Lewis defends Humeanism by refining the Principal Principle. The paper also includes an overview of Humeanism and anti-Humeanism, and critique of Anti-Humeanism.

Loewer, Barry. (2012). Two accounts of laws and time. *Philosophical Studies, 160*(1), 115–137.

This paper supports Humeanism by arguing that it can explain the operation of the laws of physics and the arrow of time if we properly distinguish between scientific and metaphysical explanations. The paper includes a presentation of prominent Humean and non-Humean views.

Skyrms, Brian. (2000). *Choice and chance: an introduction to inductive logic.* Belmont, CA: Wadsworth, Inc.

An introduction to many issues in philosophy of probability, including probability calculus, distinction between kinds of probabilities, and scientific induction.

Part VI: Are Sexes Natural Kinds?

Ayala, Saray, & Vasilyeva, Nadya. (2015). Extended sex: an account of sex for a more just society. *Hypatia, 30,* 725–742.

Ayala and Vasilyeva argue that sex categorizations based on the male/female dichotomy are inaccurate, incoherent, harmful, and oppressive. In their alternative conception of sex, sex categorization depends on properties of ourselves that we control and are not fixed.

Butler, Judith. (1990). *Gender trouble: feminism and the subversion of identity.* London: Routledge.

Butler argues that sex is a social construct, i.e. that it is not a biological natural kind, and rejects the distinction between sex and gender.

Bird, Alexander, & Tobin, Emma. (2017). Natural kinds. In Edward N. Zalta (Ed.), *The Stanford Encyclopedia of Philosophy (Spring 2017 Edition).* URL = <https://plato.stanford.edu/archives/spr2017/entries/natural-kinds/>

An overview of philosophical discussions of natural kinds from the perspective of metaphysics, philosophy of language, and philosophy of science.

Laqueur, Thomas. (1990). *Making sex: body and gender from the Greeks to Freud.* Cambridge, MA: Harvard University Press.

An account of what sex is from a historical perspective. Laqueur reviews views of sex in the western world from ancient Greece to Freud, and concludes that sex is an artifice.

Index